NIST Special Publication SP250-70

Specular Gloss

Maria E. Nadal
Edward A. Early*
E. Ambler Thompson

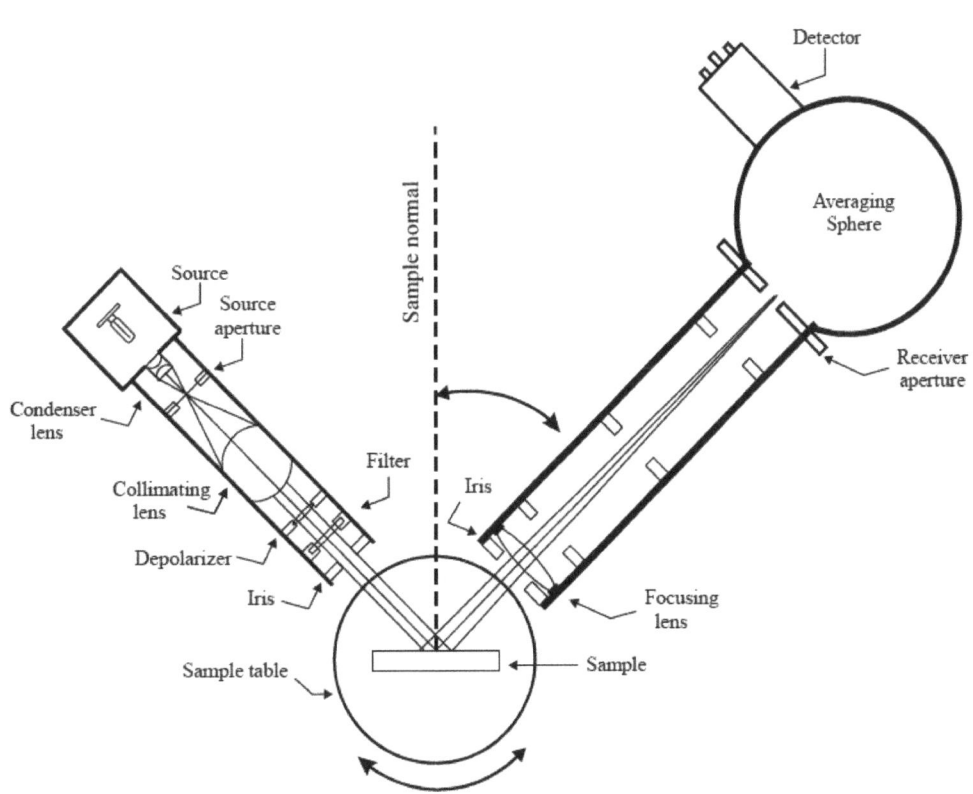

NIST Special Publication SP250-70

Specular Gloss

Maria E. Nadal
Edward A. Early*

*Optical Technology Division
Physics Laboratory
National Institute of Standards and Technology
Gaithersburg, MD 20899-0001*

E. Ambler Thompson

*Weights and Measures
Technology Services Laboratory
National Institute of Standard and Technology
Gaithersburg, MD 20899-001*

August 2006

U.S. Department of Commerce
Carlos M. Gutierrez, Secretary

Technology Administration
Robert Cresanti, Under Secretary for Technology

National Institute of Standards and Technology
William Jeffrey, Director

*AFRL/HEDO, 2650 Loiuse Bauer Drive, Brooks City-Base, TX 78235

Certain commercial entities, equipment, or materials may be identified in this document in order to describe an experimental procedure or concept adequately. Such identification is not intended to imply recommendation or endorsement by the National Institute of Standards and Technology, nor is it intended to imply that the entities, materials, or equipment are necessarily the best available for the purpose.

PREFACE

The calibration and related measurement services of the National Institute of Standards and Technology (NIST) are intended to assist the makers and users of precision measuring instrument in achieving the highest possible levels of accuracy, quality, and productivity. NIST offers over 300 different calibration, special test, and measurement assurance services. These services allow customers to directly link their measurement systems to measurement systems and standards maintained by NIST. These services are offered to the public and private organizations alike. They are described in NIST Special Publication (SP) 250, NIST Calibration Services Users Guide.

The User Guide is supplemented by a number of Special Publications (designated as the "SP250 Series") that provide detailed descriptions of the important features of specific NIST calibration services. These documents provide a description of the (1) specifications for the services; (2) design philosophy and theory; (3) NIST measurement services; (4) NIST operational procedures; (5) assessment of the measurement uncertainty including random and systematic errors and an error budget; and (6) internal quality control procedures used by NIST. These documents will present more detail than can be given in NIST calibration reports, or is generally allowed in articles in scientific journals. In the past, NIST has published such information in a variety of ways. This series will make this type of information more readily available to the user.

This document, SP250-70 (2006), NIST Measurement Services: Specular Gloss, describes the instrumentation, standards, and techniques used to measure the specular gloss of materials at 20°, 60°, and 85° geometries. A description of Special Test 38090S is also presented. Inquires concerning the technical content of this document or the specifications for these services should be directed to the authors or to one of the technical contacts cited in SP 250.

NIST welcomes suggestions on how publications such as this might be made more useful. Suggestions are also welcome concerning the need for new calibration services, special tests, and measurement assurance programs.

Belinda L. Collins
Director
Technology Services

Katharine B. Gebbie
Director
Physics Laboratory

ABSTRACT

This document describes the instrumentation, standards, and measurement techniques used at the National Institute of Standards and Technology (NIST) to measure specular gloss. The organization of this document is as follow. Section 1 describes the motivation for establishing a reference goniophotometer and primary standards for specular gloss and associated calibration services that are available. Section 2 presents the theory and measurement equations relevant to the measurements described in this document. The NIST reference goniophotometer including the illuminator, goniometer, receiver, and the characterization and uncertainty analysis of the instrument are described in Sec. 3. In Sec. 4 the new primary specular gloss standard and its characterization are presented. A sample calibration report and references containing details of the instrument and primary standard are reproduced in the appendices.

Key words: appearance attributes; gloss standard; glossmeter; goniophotometer; luminous reflectance; specular gloss

TABLE OF CONTENTS

Abstract v

1. Introduction 1

2. Theory 2
   2.1. Definitions 2
   2.2 Measurement Equations 3

3. Reference Goniophotometer 7
   3.1 Description 7
       3.1.1 Illuminator 8
       3.1.2 Goniometer 9
       3.1.3 Receiver 9
   3.2 Characterization 9
       3.2.1 Illuminator 10
       3.2.2 Goniometer 11
       3.2.3 Receiver 11
   3.3 Operation 14
   3.4 Uncertainty Analysis 14
       3.4.1 Basic Definitions 14
       3.4.2 Specular Gloss 15

4. Primary Gloss Standard 18
   4.1 Description 18
   4.2 Characterization 19
       4.2.1 Index of Refraction 20
       4.2.2 Luminous Reflectance 21
   4.3 Comparison of Specular Gloss Standards 22

Acknowledgments 23

References 24

Appendix A: Sample of Report of Calibration for Specular Gloss at 20°, 60°, and 85° geometries
Appendix B: NIST Reference Goniophotometer for Geometrical Appearance Measurements (copy of Ref. 9)
Appendix C: New Primary Standard for Specular Gloss Measurements (copy of Ref. 21)

LIST OF TABLES

Table 2.1	Geometries for specular gloss measurements and applications	2
Table 2.2	Specular reflectance $\rho_0(\theta, \lambda_D)$ of the theoretical gloss standard for each illumination angle of the standard geometries at wavelength $\lambda_D = 589.3$ nm	5
Table 3.1	ASTM specifications and the NIST measured values for illuminator and receiver apertures with tolerances and uncertainties	9
Table 3.2	Average gain ratio and relative standard deviation for successive amplifier gain ratios	12
Table 3.3	Correction factors for deviations from the standard conditions for a black glass calibrated with the primary specular gloss standard	16
Table 3.4	a. Components of uncertainty that depend on the scattering properties of the materials and the resulting relative standard uncertainties. The values are based on a BaK50 primary gloss standard and a highly polished black glass test sample	17
	b. Components of uncertainty that are independent of the scattering properties of the materials and the resulting relative standard uncertainties	17
Table 3.5	Relative expanded uncertainties of specular gloss measured by the NIST reference goniophotometer ($k = 2$) for the 20°, 60°, and 85° geometries	18
Table 4.1	Index of refraction n of BaK50 glass as a function of wavelength	20
Table 4.2	Calculated specular reflectance $\rho_s(\theta, \lambda_D)$ and specular gloss value $G_s(\theta)$ of the new primary standard at $\lambda_D = 589.3$ nm for each standard geometry	21
Table 4.3	Average luminous reflectance $\rho_{v,s}(\theta)$ and specular gloss value $G_s(\theta)$ for the new primary standard for each standard geometry and method of calculation or measurement	21

LIST OF FIGURES

Figure 3.1	Schematic of the NIST reference goniophotometer	8
Figure 3.2	Spectral flux distribution of the CIE standard illuminant C and the CIE spectral luminous efficiency function	11
Figure 3.3	Relative responsivity (ratio of the measured signal to the actual flux) of the photodiode-amplifier combination as a function of current for the listed gain settings	13
Figure 3.4	Actual and ideal spectral products of the illuminator and receiver	13
Figure 4.1	Schematic diagram of the new NIST primary standard showing the incoming and reflected beam at a specular geometry	19
Figure 4.2	Fitted refractive index, $n(\lambda)$, for BaK50 glass as a function of wavelength	20
Figure 4.3	Fitted refractive index, $n(\lambda)$, for average black glass working standards, NIST old standard (Carrara black glass), BaK50, and quartz as a function of wavelength	23
Figure 4.4	Normalized refractive index for BaK50, average of black glass working standards, and quartz as a function of wavelength	23

1. Introduction

This document describes the NIST reference goniophotometer and primary standards, as they existed at the time of publication, for the calibration of specular gloss at the 20°, 60°, and 85° standard geometries. Updated information about calibration services is published periodically in the NIST Calibration Services Users Guide [1].

Specular gloss is the perception by an observer of the mirror-like appearance of a surface [2]. The measurement of specular gloss consists of comparing the luminous reflectance from a test sample to that from a calibrated gloss standard, under the same experimental conditions. Therefore, gloss is a dimensionless and psychophysical quantity whose accurate determination depends on the characteristics of the measuring instrument and on the gloss standard. Several documentary standards describe the proper measurement conditions to determine specular gloss for specific surfaces. In particular, the International Organization for Standards ISO 2813 [3] and the American Society for Testing Materials ASTM D523 [4] describe the measurement procedure for specular gloss of nonmetallic samples. These documentary standards specify the spectral and geometrical conditions of measurement. The spectral flux distribution of the illuminator is CIE standard illuminant C and the spectral responsivity of the receiver is the CIE spectral luminous efficiency function [5]. There are three standard geometries, corresponding to illumination angles of 20°, 60°, and 85°, each with specifications on the angular extent of the rays within the influx and efflux.

The NIST reference goniophotometer was originally developed at NIST in the 1970's. In 1999, this instrument was updated, automated, and characterized to ensure proper operation and to verify agreement with the ISO and ASTM standards for specular gloss of nonmetallic samples.

The accuracy of specular gloss measurements depends not only on the properties of the instrument but also to a considerable extent on the primary gloss standard. A new primary gloss standard using BaK50 barium crown glass has been developed at NIST. It possesses high chemical and mechanical durability and an index of refraction at the sodium D line of $n_D = 1.5677$. Different calibration procedures are detailed, and the new standard is compared with other primary standards. The new gloss standard and the NIST reference goniophotometer provide an accurate calibration facility for specular gloss. Note that the primary standard referenced in the documentary standards is referred to here as the theoretical standard, while the realization of the theoretical standard is referred to here as the primary standard.

The NIST Calibration Services Program offers Service ID Number 38090S, Special Test of Specular Gloss, as listed in the Optical Properties of Materials section of the Optical Radiation Measurements Chapter of the NIST Calibration Services Users Guide [1]. The measurement service quality system is based on the ISO 17025 standard.

2. Theory

This section details the basic definitions and the relevant measurement equations to determine specular gloss. The approach used in this paper is based upon the concepts presented in Refs. [6-9].

2.1 Definitions

Gloss is the perception by an observer of the shiny appearance of a surface. This perception changes whenever there is a change in the relative position or spectral distribution of the source, the sample, or the observer. Different geometries are used to determine the specular gloss of materials, as shown in Table 2.1. These geometries were selected based on their ability to produce optimum discrimination between samples and to correlate with visual rankings. Several documentary standards specify the geometrical and spectral conditions of measurement to determine specular gloss for specific surfaces.

Table 2.1. Geometries for specular gloss measurements and applications

Illumination angle	Applications
20°	High gloss of plastic film, appliance and automotive finishes
30°	High gloss of image-reflecting surfaces
45°	Porcelain enamels and plastics
60°	All ranges of gloss for paint and plastics
75°	Coated waxes and paper
85°	Low gloss of flat matte paints and camouflage coatings

The measurement of specular gloss compares the specular luminous reflectance from a test specimen to that from a standard surface under the same experimental conditions. The illumination beam is collimated with a spectral flux distribution approximating that of CIE standard illuminant C. The luminous flux in the reflected beam is measured with a receiver having a spectral responsivity that approximates the CIE spectral luminous efficiency function V_λ. The theoretical standard for specular gloss measurements is a highly polished plane black glass with a refractive index for the sodium D line of $n_D = 1.567$. To set the specular gloss scale, the specular gloss of this theoretical standard has an assigned value of 100 for each of the three standard specular geometries of 20°, 60°, and 85°. For primary standards with a refractive index different from $n_D = 1.567$, the specular gloss is computed from the refractive index and the Fresnel equations.

2.2 Measurement Equations

The NIST reference goniophotometer measures luminous flux Φ_v [lm], defined by

$$\Phi_v = K_M \cdot \int \Phi(\lambda) \cdot V_\lambda(\lambda) \cdot d\lambda , \qquad (2.1)$$

where $K_M = 683$ lm/W is the maximum spectral luminous efficacy for photopic vision, λ [nm] is the wavelength, $\Phi(\lambda)$ [W/nm] is the spectral flux, and $V_\lambda(\lambda)$ is the spectral luminous efficiency function. For specular gloss measurements, the documentary standards specify that $\Phi(\lambda)$ have the spectral flux distribution of CIE standard illuminant C, so that the spectral flux is $\Phi_C(\lambda)$. The goniophotometer has a spectral flux $\Phi(\lambda)$ and responsivity $V(\lambda)$ that closely approximate the specified $\Phi_C(\lambda)$ and $V_\lambda(\lambda)$, respectively.

The specular luminous reflectance $\rho_v(\theta_0)$ of an object at the standard illumination angle θ_0 (20°, 60°, or 85°) is given by

$$\rho_v(\theta_0) = \frac{\Phi_{v,r}(\theta_0)}{\Phi_{v,i}} , \qquad (2.2)$$

where $\Phi_{v,r}(\theta_0)$ and $\Phi_{v,i}$ are the reflected and incident luminous fluxes, respectively. In terms of the geometrical and spectral properties of the instrument and object, the luminous fluxes are given by

$$\Phi_{v,r}(\theta_0) = \int d\theta \int d\lambda \cdot \Phi_{C,i}(\theta, \lambda) \cdot \rho(\theta, \lambda) \cdot V_\lambda(\lambda) \text{ and} \qquad (2.3)$$

$$\Phi_{v,i} = \int d\lambda \cdot \Phi_{C,i}(\lambda) \cdot V_\lambda(\lambda) , \qquad (2.4)$$

where $\rho(\theta, \lambda)$ is the specular spectral reflectance of the object. The angles θ are the angles around θ_0 resulting from the finite size of the source aperture. There is no angular dependence for $\Phi_{v,i}$ since the entire incident beam is collected by the receiver.

The current I [A] from the detector of the receiver is given by

$$\begin{aligned}
I &= \int \Phi(\lambda) \cdot R(\lambda) \cdot d\lambda \\
&= \int \Phi(\lambda) \cdot R \cdot V_\lambda(\lambda) \cdot r(\lambda) \cdot d\lambda , \\
&\cong \frac{R}{K_M} \cdot \Phi_v
\end{aligned} \qquad (2.5)$$

where R [A/W] is the responsivity of the detector and $r(\lambda)$ is the difference between the spectral luminous efficiency function and the actual spectral responsivity of the detector. The current is converted to a voltage N [V], given by

$$N = I \cdot A , \qquad (2.6)$$

where A [V/A] is the amplifier gain. Solving Eq. (2.5) for the luminous flux and substituting into Eq. (2.2) yields

$$\rho_v(\theta_0) = \frac{N_r}{N_i} \cdot \frac{A_i}{A_r}, \qquad (2.7)$$

where the subscripts i and r refer to measurements made of the incident and reflected luminous fluxes, respectively.

The specular gloss of a test sample at illumination angle θ_0, $G_t(\theta_0)$, is given by

$$G_t(\theta_0) = G_s(\theta_0) \cdot \frac{\rho_{v,t}(\theta_0)}{\rho_{v,s}(\theta_0)}, \qquad (2.8)$$

where $G_s(\theta_0)$ is the specular gloss of the primary standard and $\rho_{v,t}(\theta_0)$ and $\rho_{v,s}(\theta_0)$ are the specular luminous reflectances of the test sample and primary standard, respectively. Substituting Eq. (2.7) for the luminous reflectances of the test sample and primary standard into Eq. (2.8) yields

$$G_t(\theta_0) = G_s(\theta_0) \cdot \frac{N_{r,t}}{N_{r,s}} \cdot \frac{A_{r,s}}{A_{r,t}}, \qquad (2.9)$$

which relates the specular gloss of the test sample to the specular gloss of the primary standard and the measured signals from the reflected luminous fluxes from the test sample and primary standard. The specular gloss of the primary standard is given by

$$G_s(\theta_0) = G_0(\theta_0) \cdot \frac{\rho_s}{\rho_0(\theta_0, \lambda_D)}, \qquad (2.10)$$

where $G_0(\theta_0)$ is the specular gloss of the theoretical standard, $\rho_0(\theta_0, \lambda_D)$ is the specular reflectance of the theoretical standard at a wavelength $\lambda_D = 589.3$ nm, and ρ_s is the specular reflectance of the primary standard. For each angle of illumination, the specular gloss of the theoretical standard is defined as $G_0(\theta) = 100$.

The specular reflectance ρ from the surface of a dielectric sample depends on the incident angle θ defined relative to the normal of the sample, the wavelength λ, and the polarization σ (p or s) of the incident radiant flux. The specular reflectance as a function of these variables is given by the Fresnel equations,

$$\rho(\theta, \lambda, \mathrm{p}) = \left[\frac{n^2(\lambda) \cdot \cos\theta - \sqrt{n^2(\lambda) - \sin^2\theta}}{n^2(\lambda) \cdot \cos\theta + \sqrt{n^2(\lambda) - \sin^2\theta}} \right]^2 \text{ and} \qquad (2.11)$$

$$\rho(\theta, \lambda, s) = \left[\frac{\cos\theta - \sqrt{n^2(\lambda) - \sin^2\theta}}{\cos\theta + \sqrt{n^2(\lambda) - \sin^2\theta}} \right]^2 , \qquad (2.12)$$

where n is the index of refraction. The specular reflectance for unpolarized incident radiant flux is calculated from

$$\rho(\theta, \lambda) = \frac{1}{2}[\rho(\theta, \lambda, p) + \rho(\theta, \lambda, s)] . \qquad (2.13)$$

From the documentary standards, the index of refraction of the theoretical standard at wavelength λ_D is $n(\lambda_D) = 1.567$. The specular reflectances at the standard angles of illumination for unpolarized radiant flux for the theoretical standard, $\rho_0(\theta_0, \lambda_D)$, are calculated from Eqs. (2.11) to (2.13) and are listed in Table 2.2

Table 2.2. Specular reflectance $\rho_0(\theta, \lambda_D)$ of the theoretical gloss standard for each standard illumination angle at wavelength $\lambda_D = 589.3$ nm

Illumination angle	$\rho_0(\theta, \lambda_D)$
20°	0.049078
60°	0.100056
85°	0.619148

The specular gloss of the primary standard depends, from Eq. (2.10), on its reflectance ρ_s. However, the documentary standards are ambiguous concerning the definition of this reflectance. One approach is to use the calculated $\rho_s(\theta_0, \lambda_D)$ of the primary standard for its index of refraction at $\lambda_D = 589.3$ nm.

An alternative approach, and the one taken for measurements with the reference goniophotometer, is to measure the luminous reflectance $\rho_{v,s}(\theta_0)$ of the primary standard. This approach has the advantages of taking the dispersion characteristics of the standard into account and being a measurable quantity on the goniophotometer. In terms of the geometrical and spectral conditions of measurement, the luminous reflectance is obtained from Eqs. (2.2) to (2.4) as

$$\rho_{v,s}(\theta_0) = \frac{\int d\theta \int d\lambda \cdot \Phi_{C,i}(\theta, \lambda) \cdot \rho_s(\theta, \lambda) \cdot V_\lambda(\lambda)}{\int d\lambda \cdot \Phi_{C,i}(\theta, \lambda) \cdot V_\lambda(\lambda)} , \qquad (2.14)$$

while in terms of the measured signals the luminous reflectance is obtained from Eq. (2.7) as

$$\rho_{v,s}(\theta_0) = \frac{N_{r,s}}{N_i} \cdot \frac{A_i}{A_{r,s}} . \qquad (2.15)$$

Using the luminous reflectance of the primary standard and Eq. (2.10), Eq. (2.9) becomes

$$G_t(\theta_0) = 100 \cdot \frac{\rho_{v,s}(\theta_0)}{\rho_0(\theta_0, \lambda_D)} \cdot \frac{N_{r,t}}{N_{r,s}} \cdot \frac{A_{r,s}}{A_{r,t}}, \qquad (2.16)$$

where $\rho_0(\theta_0, \lambda_D)$ is calculated and given in Table 2.2, $\rho_{v,s}(\theta_0)$ is measured and given by Eq. (2.15), and the signals $N_{r,t}$ and $N_{r,s}$ are the measured signals from the test sample and primary standard, respectively, at amplifier gains of $A_{r,t}$ and $A_{r,s}$. Equation (2.16) is the measurement equation for specular gloss in terms of measured and calculated quantities.

In terms of the spectral and geometrical quantities, using the appropriate modification of Eq. (2.14) in Eqs. (2.8) and (2.10), the specular gloss of a test samples is given by

$$G_t(\theta_0) = 100 \cdot \frac{\rho_{v,s}(\theta_0)}{\rho_0(\theta_0, \lambda_D)} \cdot \frac{\int d\theta \int d\lambda \cdot \Phi_i(\theta, \lambda) \cdot \rho_t(\theta, \lambda) \cdot V(\lambda)}{\int d\theta \int d\lambda \cdot \Phi_i(\theta, \lambda) \cdot \rho_s(\theta, \lambda) \cdot V(\lambda)}, \qquad (2.17)$$

where $\Phi_i(\theta, \lambda)$ and $V(\lambda)$ are the incident spectral flux and responsivity of the reference instrument, respectively. In terms of the ideal case given in the documentary standards, the specular gloss $G_{t,id}$ is given by

$$G_{t,id}(\theta_0) = 100 \cdot \frac{\rho_{v,s}(\theta_0)}{\rho_0(\theta_0, \lambda_D)} \cdot \frac{\int d\theta \int d\lambda \cdot \Phi_{C,i}(\theta, \lambda) \cdot \rho_t(\theta, \lambda) \cdot V_\lambda(\lambda)}{\int d\theta \int d\lambda \cdot \Phi_{C,i}(\theta, \lambda) \cdot \rho_s(\theta, \lambda) \cdot V_\lambda(\lambda)}, \qquad (2.18)$$

Deviations of the reference instrument from the ideal case include deviations from the specified spectral flux distribution of the illuminator, spectral responsivity of the receiver, unpolarized and collimated illumination beam with specified angular ranges, and illumination angle. The correction factors for the specular gloss of the sample under test as determined with the reference instrument are obtained using Eqs. (2.17) and (2.18). The correction factors are the ratio of the specular gloss for the ideal case to the specular gloss for the reference instrument,

$$C = \frac{G_{t,id}(\theta_0)}{G_t(\theta_0)}. \qquad (2.19)$$

The correction factor for deviations from the ideal spectral flux distribution of the illuminator and spectral responsivity of the receiver is given by

$$C_s = \frac{\int d\lambda \cdot \Phi_{C,i}(\lambda) \cdot \rho_t(\theta_0, \lambda) \cdot V_\lambda(\lambda)}{\int d\lambda \cdot \Phi_{C,i}(\lambda) \cdot \rho_s(\theta_0, \lambda) \cdot V_\lambda(\lambda)} \cdot \frac{\int d\lambda \cdot \Phi_i(\lambda) \cdot \rho_s(\theta_0, \lambda) \cdot V(\lambda)}{\int d\lambda \cdot \Phi_i(\lambda) \cdot \rho_t(\theta_0, \lambda) \cdot V(\lambda)}. \qquad (2.20)$$

The correction factor for deviations from an unpolarized illumination beam is given by

$$C_{\mathrm{p}} = \frac{\int d\lambda \cdot \Phi_{\mathrm{C,i}}(\lambda) \cdot \rho_{\mathrm{t}}(\theta_0,\lambda) \cdot V_\lambda(\lambda)}{\int d\lambda \cdot \Phi_{\mathrm{C,i}}(\lambda) \cdot \rho_{\mathrm{s}}(\theta_0,\lambda) \cdot V_\lambda(\lambda)} \cdot \frac{\int d\lambda \cdot \Phi_{\mathrm{C,i}}(\lambda) \cdot \rho_{\mathrm{s}}(\theta_0,\lambda,\sigma) \cdot V_\lambda(\lambda)}{\int d\lambda \cdot \Phi_{\mathrm{C,i}}(\lambda) \cdot \rho_{\mathrm{t}}(\theta_0,\lambda,\sigma) \cdot V_\lambda(\lambda)} . \qquad (2.21)$$

The correction factor for deviations from the angular ranges of the illumination beam is given by

$$C_{\mathrm{d}} = \frac{\int d\theta_{\mathrm{id}} \int d\lambda \cdot \Phi_{\mathrm{C,i}}(\theta,\lambda) \cdot \rho_{\mathrm{t}}(\theta,\lambda) \cdot V_\lambda(\lambda)}{\int d\theta_{\mathrm{id}} \int d\lambda \cdot \Phi_{\mathrm{C,i}}(\theta,\lambda) \cdot \rho_{\mathrm{s}}(\theta,\lambda) \cdot V_\lambda(\lambda)} \cdot \frac{\int d\theta \int d\lambda \cdot \Phi_{\mathrm{C,i}}(\theta,\lambda) \cdot \rho_{\mathrm{s}}(\theta,\lambda) \cdot V_\lambda(\lambda)}{\int d\theta \int d\lambda \cdot \Phi_{\mathrm{C,i}}(\theta,\lambda) \cdot \rho_{\mathrm{t}}(\theta,\lambda) \cdot V_\lambda(\lambda)}$$

(2.22)

The correction factor for deviations from the standard illumination angle is given by

$$C_{\mathrm{a}} = \frac{\int d\lambda \cdot \Phi_{\mathrm{C,i}}(\lambda) \cdot \rho_{\mathrm{t}}(\theta_0,\lambda) \cdot V_\lambda(\lambda)}{\int d\lambda \cdot \Phi_{\mathrm{C,i}}(\lambda) \cdot \rho_{\mathrm{s}}(\theta_0,\lambda) \cdot V_\lambda(\lambda)} \cdot \frac{\int d\lambda \cdot \Phi_{\mathrm{C,i}}(\lambda) \cdot \rho_{\mathrm{s}}(\theta,\lambda) \cdot V_\lambda(\lambda)}{\int d\lambda \cdot \Phi_{\mathrm{C,i}}(\lambda) \cdot \rho_{\mathrm{t}}(\theta,\lambda) \cdot V_\lambda(\lambda)} . \qquad (2.23)$$

Therefore, the specular gloss for the ideal case in terms of the correction factors is given by

$$G_{\mathrm{t,id}}(\theta_0) = G_{\mathrm{t}} \cdot C_{\mathrm{s}} \cdot C_{\mathrm{p}} \cdot C_{\mathrm{d}} \cdot C_{\mathrm{a}} . \qquad (2.24)$$

The final measurement equation, including the correction factors and the experimentally measured quantities, is given by

$$G_{\mathrm{t,id}}(\theta_0) = 100 \cdot \frac{\rho_{\mathrm{v,s}}(\theta_0)}{\rho_0(\theta_0,\lambda_{\mathrm{D}})} \cdot \frac{N_{\mathrm{t}}}{N_{\mathrm{s}}} \cdot \frac{A_{\mathrm{s}}}{A_{\mathrm{t}}} \cdot C_{\mathrm{s}} \cdot C_{\mathrm{p}} \cdot C_{\mathrm{d}} \cdot C_{\mathrm{a}} . \qquad (2.25)$$

3. Reference Goniophotometer

3.1 Description

Interest in specular gloss measurements at NIST (formerly NBS) dates back to the 1930's [10-12]. The goniophotometer described in this paper was originally developed at NBS [6] in the 1970's by scientists active in the development of the ASTM standard test method, and has recently been updated, automated, and characterized. This instrument is a monoplane gonio-instrument with a fixed illuminator and a rotating sample Table and receiver arm. The sample Table and the receiver arm can be rotated independently of each other in the plane of incidence. A schematic diagram of the goniophotometer in reflectance mode is shown in Fig. 3.1 and a detailed description of the instrument is given below. This instrument complies with the geometrical and spectral conditions specified by the ASTM D523 and ISO 2813 documentary standards for the measurement of specular gloss of non-metallic surfaces at fixed specular geometries of 20°, 60°, and 85°.

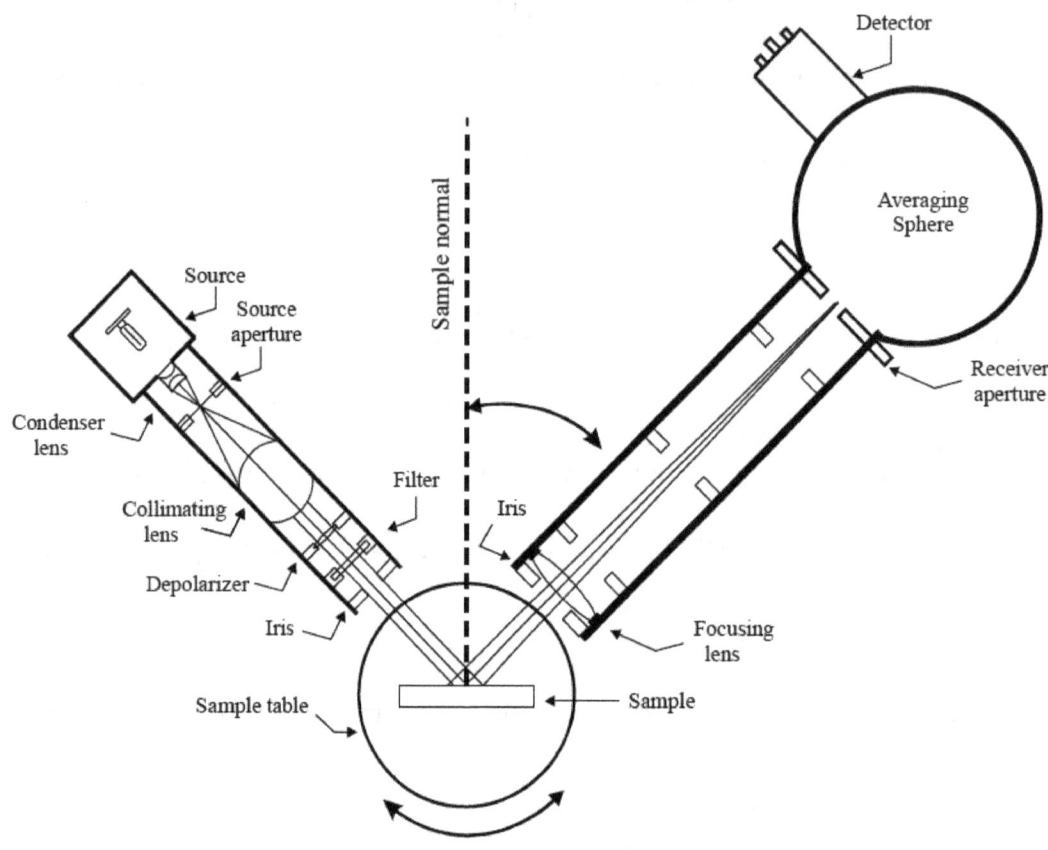

Figure 3.1. Schematic of the NIST Reference Goniophotometer

3.1.1 Illuminator The illuminator supplies the influx onto the sample. It consists of a light source, a set of condenser lenses, a source aperture, a collimating lens, a depolarizer, a color filter, and an iris. The light source is a quartz-tungsten-halogen incandescent lamp rated at 100 W. A constant dc current of 6.3 A is run through the lamp from a computer-controlled power supply. This current was chosen so that the spectral flux distribution from the lamp approximates CIE standard illuminant A. An image of the lamp filament is formed at the source aperture with the achromatic condenser lens system. This source aperture is the field stop of the illuminator. The recommended and actual dimensions of the source aperture are given in Table 3.1. The source aperture is at the focus of an achromatic lens with a focal length of 192 mm, which collimates the beam. The polarization of the beam is determined using either a glass-laminated polarizer to provide linearly polarized light, or a scrambler to provide unpolarized light. The scrambler is used for most measurements. A color filter BG-34, 2.7 mm thick, converts the spectral flux distribution to CIE standard illuminant C. Finally, an iris diaphragm serves as the aperture stop of the illuminator. For most measurements the beam diameter is about 3 cm.

Table 3.1. ASTM specifications and the NIST measured values for full angles of the illuminator and receiver apertures with tolerances and uncertainties

Apertures	Parallel to the plane of incidence		Perpendicular to the plane of incidence	
	ASTM spec.	NIST apertures	ASTM spec.	NIST apertures
Source	0.75° ± 0.25°	0.71° ± 0.01°	2.5° ± 0.5°	3.0° ± 0.01°
20° receiver	1.8° ± 0.05°	1.82° ± 0.01°	3.6° ± 0.1°	3.64° ± 0.02°
60° receiver	4.4° ± 0.1°	4.44° ± 0.02°	11.7° ± 0.2°	11.72° ± 0.01°
85° receiver	4.0° ± 0.3°	4.03° ± 0.01°	6.0° ± 0.3°	6.02° ± 0.01°

3.1.2 Goniometer The goniometer positions the sample and receiver for bi-directional luminous reflectance or transmittance measurements. It consists of two computer-controlled rotation stages. The sample stage has a diameter of 48 cm and the sample holder can accommodate rectangular samples with dimensions up to 10 cm high by 20 cm wide by 25 mm thick. In addition, a sample holder with vacuum suction can be mounted on the sample stage for use with non-rigid samples such as paper. The sample holder positions the front surface of the sample on the axis of rotation of the stage. The receiver is attached to a rotation stage whose axis of rotation is collinear with the rotation stage for the sample.

3.1.3 Receiver The receiver collects and measures the efflux from the sample. It consists of an iris, a focusing lens, a receiver aperture, an averaging sphere, and a detector. The iris, lens, and aperture are enclosed in a light-tight tube with baffles to reduce scattered light. The iris has a clear diameter of 7.4 cm and is the aperture stop of the receiver. A lens with a focal length of 508 mm immediately behind the iris focuses the light at the receiver aperture. The specified dimensions of the receiver aperture, with tolerances, and the measured dimensions of the receiver aperture are given in Table 3.1. The receiver aperture is located at the entrance port of the averaging sphere, and is the field stop of the receiver. The averaging sphere has a diameter of 30.5 cm and is packed with pressed polytetrafluoroethylene (PTFE). A photopic detector is located at the detector port of the averaging sphere and consists of a silicon photodiode and a filter that in combination approximates the spectral luminous efficiency function V_λ. The detector is temperature controlled with a thermoelectric heater and has an integral current-to-voltage amplifier. The voltage output from the detector is measured by a 6 ½ digit voltmeter and sent to a computer.

3.2 Characterization

This instrument has been carefully characterized and calibrated to ensure proper operation and to verify agreement with the documentary standards for specular gloss measurements - ASTM D523 and ISO 2813. These documentary standards specify the geometry of the illumination beam, the specular angles, the dimensions of the source and receiver apertures, the spectral flux distribution of the illuminator, and the spectral responsivity of the receiver. Details of the characterization of the instrument are given below.

3.2.1 Illuminator The lamp is burned in at 100 W for 24 hours and then at the operational current of 6.3 A for an additional 48 h prior to performing measurements. The stability of the source was investigated by measuring the flux incident on a Si photodiode at the sample location for a period of 30 min. A typical relative standard deviation of the signal over ten minutes is 0.01 %.

The degree of polarization of the illuminator with no polarizing components was measured by rotating a polarizer in the beam and measuring the maximum and minimum transmitted signals. The difference of the signals was divided by the sum to obtain the degree of polarization of the source, approximately 9 %. Because the illuminator is not completely unpolarized, either a glass-laminated polarizer or a scrambler are used in the beam path just before the sample. The averaging sphere on the receiver arm in front of the photopic detector randomizes any additional polarization imparted by the sample. The leakage of the glass-laminated polarizer for luminous flux that is polarized perpendicular to the intended direction of polarization of the illuminator is 1.4 %. The degree of polarization of the source-scrambler combination is 0.02 %, providing an unpolarized illumination beam.

The standard spectral conditions for specular gloss measurements are for the illuminator to approximate the spectral flux distribution of CIE standard illuminant C and for the receiver spectral responsivity to approximate the CIE spectral luminous efficiency function, both shown in Fig. 3.2. The spectral flux distribution of the illuminator, with no filter, for the spectral range of 380 nm to 1070 nm was measured using a calibrated scanning spectroradiometer [13]. A freshly pressed polytetrafluoroethylene (PTFE) sample [14] was placed in the sample holder at an illumination angle of 45° and a viewing angle of approximately 0°. The spectral reflectance of PTFE is non-selective in the visible region. The lamp current was set to 6.3 A to achieve a Correlated Color Temperature (CCT) of 2856 K ± 10 K with a coverage factor $k = 2$. A color-temperature conversion filter was placed in the illuminator to achieve the spectral flux distribution of CIE standard illuminant C (CCT = 6774 K). The spectral flux distribution of the illuminator was measured as before, and a CCT of 6740 K ± 30 K with a coverage factor $k = 2$ was calculated.

The dimensions of the source and receiver apertures are predetermined by the specular geometry, focal length of the collimator lens in the system, test material, and the correlation with visual ranking. Table 3.1 lists the standard aperture sizes and accepTable tolerances, and the measured values for the NIST apertures. The linear dimensions of the apertures were measured and the angular dimensions were calculated. All of the apertures are within the ASTM tolerances. The angle of incidence θ for any ray within the illumination beam with illumination angle θ_0 is given by

$$\cos\theta = \frac{\cos\theta_0 + \sin\theta_0 \cdot \tan\alpha}{(1+\tan^2\alpha+\tan^2\beta)^{1/2}} \quad , \tag{3.1}$$

where α and β are the angles from the illuminator axis parallel and perpendicular to the plane of incidence, respectively.

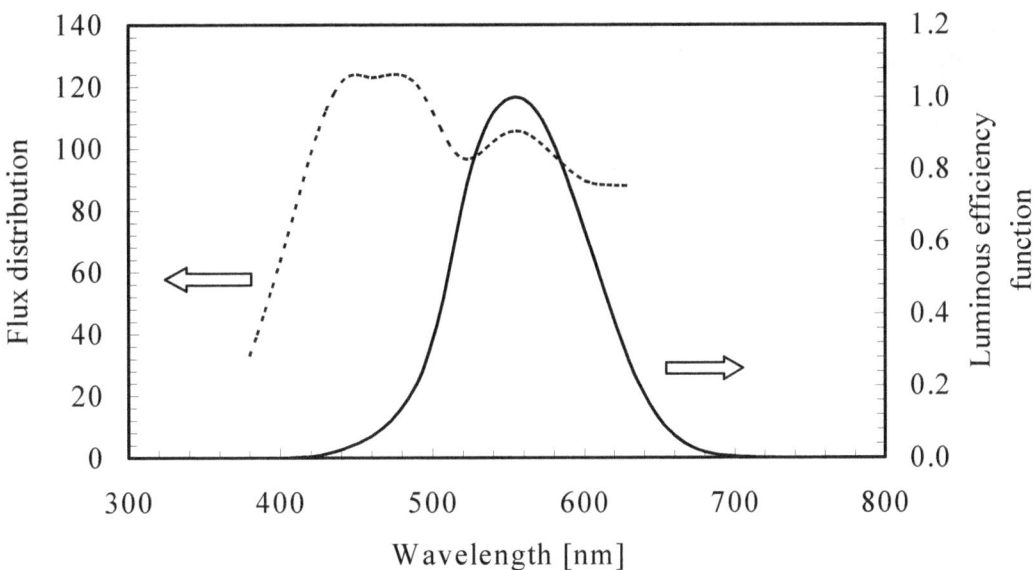

Figure 3.2. Spectral distributions of CIE standard illuminant C (dashed line) and the CIE spectral luminous efficiency function (solid line)

3.2.2 Goniometer The angular scales of the sample Table and receiver were checked for accuracy and repeatability from 0° to ± 85°. The illumination angles on the sample Table were calibrated using a set of isosceles prisms made by the NIST Optical Shop. These prisms have nominal base angles of 20°, 60°, and 85° with maximum deviations of 0.05°, corresponding to the standard geometries. The prisms were calibrated by the Manufacturing Engineering Laboratory at NIST using a precision electronic autocollimator with an expanded uncertainty (k=2) of 0.005° [15]. For the calibration of the illumination angles, one of the calibrated prisms was mounted in the sample holder. The sample Table was then rotated to achieve retroreflection. This procedure was repeated for the three prisms, resulting in a maximum uncertainty for the angular scale of the sample Table of 0.05°. The repeatability of the angular scale for the sample Table is 0.05°. To calibrate the receiver angular scale, a front surface mirror replaced the prism and the reflection was centered in the receiver aperture by rotating the receiver arm. The maximum uncertainty for the receiver arm angular scale is 0.05° and the repeatability is 0.05°.

3.2.3 Receiver The photopic receiver consists of a temperature controlled silicon photodiode with a photopic filter and a computer controlled variable gain current-to-voltage amplifier. The gain setting of the amplifier is the power of ten by which the current is multiplied to convert it to voltage. For low gloss samples, the measured luminous flux is much lower than the luminous flux from the high specular gloss primary standard. In this case, several amplifier gain settings are required to cover the dynamic range. From eq (2.25), the ratio of the amplifier gains is the required quantity. The gain settings of the amplifier were calibrated using a reference constant current source as the input of the amplifier and a voltmeter was used to measure the output signal. At each gain setting, the current was set to obtain a range of voltages from 10 V to 0.1 V in twenty equally spaced steps and the output signal was measured at each current. This protocol yielded current settings that were common to successive gain settings. The average value of the gain ratio and relative standard deviation of these overlapping settings for successive gains are listed in Table 3.2. The gain ratios are not exactly equal to 10, and it is best to measure signals between the ranges of 0.5 V to 12 V.

Table 3.2. Average gain ratio and relative standard deviation for successive amplifier gain ratios

Gain ratio	Average value	Relative standard deviation [%]
A_6/A_5	9.999	0.002
A_7/A_6	9.998	0.002
A_8/A_7	9.997	0.002
A_9/A_8	10.017	0.003
A_{10}/A_9	9.952	0.003

The linearity of the photodiode-amplifier combination was measured using the NIST automated beam conjoiner [16]. This facility uses the beam addition method with a set of filters in the path of the two branches of the beam to vary the radiant flux at the receiver by three decades. The linearity of the photodiode-amplifier combination was checked at gain settings from 5 to 9. The maximum radiant flux at the receiver is controlled by a set of neutral-density filters in front of the receiver. Figure 3.3 shows a plot of the relative responsivity (ratio of the measured signal to the actual flux) as a function of current for the gain settings listed in the legend. The measured relative responsivity of the photodiode-amplifier combination is linear to within 0.2 % for gain settings of 5 to 8. Larger deviations are observed at gain 9 due to the small current at this setting.

The absolute spectral responsivity of the filter and receiver package was measured in the NIST Spectral Comparator Facility [17], and approximates that of the CIE spectral luminous efficiency function. The measured spectral distribution was normalized to one at 560 nm to compare with the theoretical distribution. The responsivity spatial uniformity of the detector was not determined since an averaging sphere is used in front of the detector. Figure 3.4 shows a comparison of the spectral products of the illuminator and receiver and of the CIE standard illuminant C spectral flux distribution and the spectral luminous efficiency function.

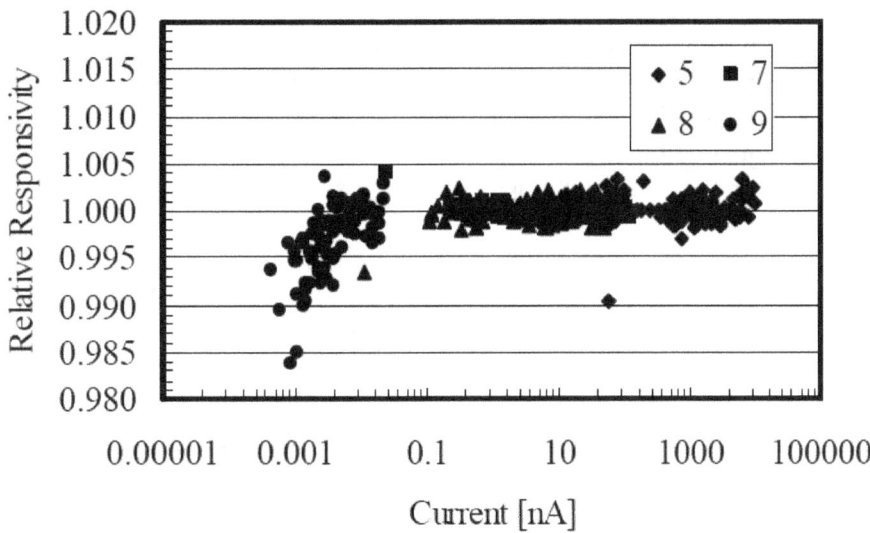

Figure 3.3. Relative responsivity (ratio of the measured signal to the actual flux) of the photodiode-amplifier combination as a function of current for the listed gain settings

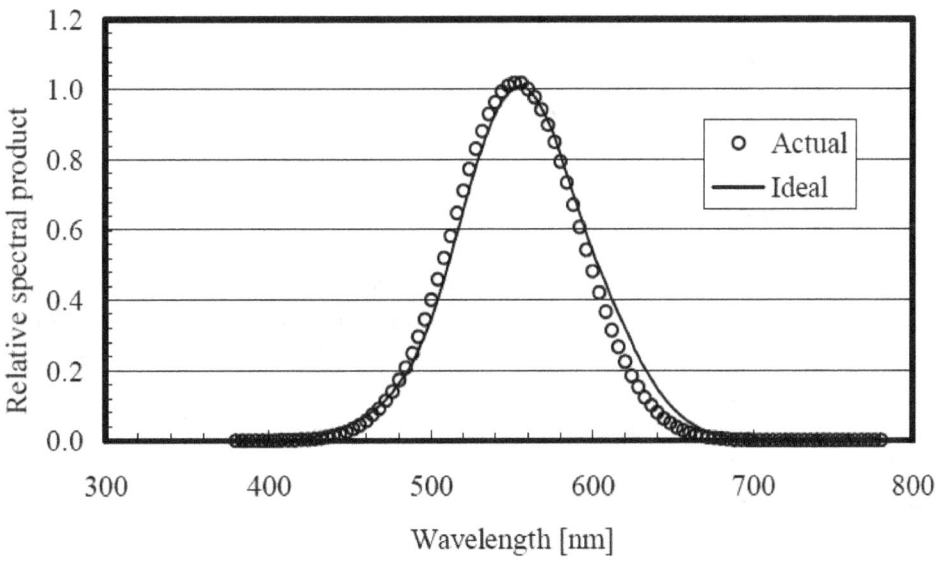

Figure 3.4. Actual and ideal spectral products of the illuminator and receiver

3.3 Operation

The lamp current is adjusted to 6.3 A and allowed to warm up for at least an hour. An image of the lamp filament is formed at the receiver aperture by adjusting the positions of the lamp and collimating lens. The zero position of the sample Table is set by placing a mirror in the sample holder and adjusting the rotation stage so that the illumination beam is retroreflected back onto the illuminator iris. The zero position of the receiver arm is adjusted by centering the illumination beam on the front of the receiver aperture. These positions are recorded by the computer program to set the rotation stages to zero. The sample holder is designed so that the front surface of the sample is on the axis of rotation of the sample Table. The sample is manually centered in the horizontal and vertical directions.

From Eq. (2.8), measurements of the specular gloss of a test sample require measurements of the reflected luminous fluxes from the test sample and the standard under the same experimental conditions. Measurements of the sample under test are bracketed with the primary standard measurements to correct for any instrumental drift. The specular gloss of the test sample is then calculated from the ratio of the measured luminous flux reflected from the sample to the average of the measured luminous fluxes reflected from the primary standard. If a polarizer is used, measurements are performed with parallel and perpendicular polarization with respect to the plane of incidence and the average of these two measurements yields the specular gloss value for unpolarized light. The dark signal is measured with the source blocked at the same geometry for the measurements, and the dark signal is subtracted from the measured signal from the standard and sample under test. The final gloss values are given by the average of at least three independent scans.

The specular gloss scale is checked before all calibrations by measuring check standards with gloss values similar to that of the sample under test. The check standards are black glass tiles with different gloss values.

3.4 Uncertainty Analysis

This section describes the components of uncertainty, their evaluation, and the resulting uncertainty for specular gloss measurements. Throughout this document, uncertainty statements follow the NIST policy given by Taylor and Kuyatt [18], which recommends the use of an expanded uncertainty with a coverage factor $k = 2$ for the uncertainties of all NIST calibrations. The NIST reference goniophotometer was designed so that the various sources of uncertainty were minimized. The residual uncertainties were characterized to produce specular gloss with a minimum relative expanded uncertainty ($k = 2$) of 0.3 %.

3.4.1 Basic Definitions
The specular gloss of a test sample is not measured directly but is determined from reflected luminous fluxes through the measurement equation. In general, the value of a measurand, y, is obtained from n other quantities x_i through a functional relation, f, given by

$$y = f(x_1, x_2, \cdots, x_i, \cdots, x_n). \tag{3.2}$$

The standard uncertainty of an input quantity is the estimated standard deviation associated with this quantity and is denoted by $u(x_i)$. The standard uncertainties may be obtained using either a

Type A evaluation of uncertainty, which is based on statistical analysis, or a Type B evaluation of uncertainty, which is based on other means. The relative standard uncertainty is given by $u(x_i)/x_i$. The relative combined standard uncertainty $u_c(y)/y$ is given by the root-sum-square of the standard uncertainties associated with each quantity x_i, assuming that these uncertainties are uncorrelated. The mathematical expression for the relative combined standard uncertainty is

$$u_c^2(y)/y^2 = \sum_{i=1}^{n} \left(\frac{1}{y} \frac{\partial f}{\partial x_i} \right)^2 u^2(x_i), \tag{3.3}$$

where $(1/y)(\partial f/\partial x_i)$ is the relative sensitivity coefficient. The expanded uncertainty U is given by $k \cdot u_c(y)$, where k is the coverage factor and is chosen on the basis of the desired level of confidence to be associated with the interval defined by U.

3.4.2 Specular Gloss The components of uncertainty associated with gloss measurements are divided into those arising from experimentally measured quantities and those arising from deviations from standard recommendations. The experimentally measured quantities include the source stability, responsivity of the detector, photodiode linearity, accuracy of the digital voltmeter, amplifier gain ratio, and signal noise. Deviations of the reference instrument from the standard conditions include deviations from the specified spectral flux distribution of the illuminator, spectral responsivity of the receiver, unpolarized influx, illumination angle, and angular distribution of the influx. The resulting uncertainty from deviations from the standard conditions depends on the nature of the sample under test and the primary standard.

The appropriate measurement equation for the first set of uncertainties is eq (2.16) and for the second set, Eqs. (2.20) to (2.23). This uncertainty analysis was applied to representative measurements using a highly polished black glass as the test sample and a highly polished wedge of BaK50 glass as the primary standard.

The components of uncertainty arising from the illuminator are uncertainties from the stability, polarization, and angular distribution of the influx. The relative standard uncertainty in source stability is caused by random effects and is 0.01 % from a Type A evaluation. The uncertainty arising from the polarization of the illuminator is caused by a systematic effect with a Type B evaluation and depends on the reflecting properties of the sample. The correction factors for deviations from an unpolarized illumination beam were calculated from Eq. (2.21) using the degrees of polarization determined in the previous section and are listed in Table 3.3 for the three standard geometries. The uncertainty arising from the angular distribution of the influx is caused by a systematic effect with a Type B evaluation and depends on the reflecting properties of the sample. The correction factor for deviations from the specified angular distribution of the influx in Table 3.1 was calculated from eqs (2.22) and (3.1). The correction factors for the three standard geometries are listed in Table 3.3.

The uncertainties arising from the goniometer are the uncertainties in the angular positioning of the sample and receiver arm. These uncertainties are caused by a systematic effect with a Type B evaluation and depend on the reflecting properties of the sample. The angular setting uncertainty of 0.05° is smaller than the accuracy of 0.1° specified by ASTM D523. The correction factors for deviations of the illumination angle from eq (2.23) for the three standard geometries are listed in Table 3.3.

The components of uncertainty arising from the receiver are uncertainties from the digital voltmeter (DVM), signal noise, amplifier gain ratio, and photodiode linearity. The uncertainty from the DVM is caused by a systematic effect with a Type B evaluation, assuming a normal probability distribution. Using the manufacturer's specifications, the relative uncertainty resulting from the DVM is 0.002 %. Signal noise of the instrument results in an uncertainty caused by a random effect with a Type A evaluation. A typical relative standard deviation for a gloss measurement is 0.01 %. The uncertainties arising from the photodiode linearity and the amplifier gain ratio are caused by systematic effects with a Type A evaluation. The maximum relative standard deviation from linearity of the photodiode, for signals in the range of 0.1 V to 12 V, is 0.1 %. The relative standard deviations for successive gain settings are listed in Table 3.2. The maximum relative standard deviation is 0.003 %. The same gain is used when the standard and sample have similar gloss values; otherwise the gain ratios are applied in the calculations.

The repeatability of the sample positioning results in an uncertainty caused by a random effect with a Type B evaluation. Assuming a rectangular probability distribution with a maximum deviation of 0.2 gloss units, the relative standard uncertainty is 0.1 %.

Deviations from the spectral condition are important for low-gloss samples with spectral features. The spectral conditions of the illuminator and receiver are in close agreement with the recommended CIE standard illuminant C and CIE spectral luminous efficiency function. The difference between the instrument and the standard conditions results in an uncertainty caused by a systematic effect with a Type B evaluation. The correction factors for deviations from the standard recommendations for the three standard geometries, from eq (2.20), are listed in Table 3.3.

Table 3.3. Correction factors for deviations from the standard conditions for a black glass calibrated with the primary specular gloss standard

Deviation	Correction factors		
	Standard geometry		
	20°	60°	85°
Polarizer, C_p	0.999890	1.000137	1.000222
Depolarizer, C_p	0.999999	1.000001	0.999995
Angular distribution, C_d	0.999999	0.999975	0.999993
Illumination angle, C_a	0.999996	0.999875	0.999940
Spectral product, C_s	1.000012	1.000006	0.999997

Since all of the correction factors listed in Table 3.3 are nearly unity, no correction factor is applied and the variations from unity are considered to be uncertainties. The relative standard uncertainty from each component of uncertainty is listed in Tables 3.4a and 3.4b for sample-dependent and sample–independent components, respectively. The relative expanded uncertainties given in Table 3.5 are calculated from the root-sum-square of the uncertainties listed in Tables 3.4a and 3.4b. The ISO 2813 standard specifies that gloss should be reported to the nearest full gloss unit. Therefore, the calibration of the reference instrument and primary standard should have an uncertainty of less than one gloss unit. The NIST reference goniophotometer has an expanded ($k = 2$) relative uncertainty of 0.3 % for calibrating a highly polished black glass.

Table 3.4a. Components of uncertainty which depend on the scattering properties of the materials, and the resulting relative standard uncertainties. The values are based on a BaK50 primary gloss standard and a highly polished black glass test sample

Component of uncertainty	Effect	Type	Relative standard uncertainty [%]		
			Standard geometry		
			20°	60°	85°
Polarizer	S	B	0.01	0.01	0.02
Depolarizer	S	B	0.0001	0.0001	0.0005
Angular distribution	S	B	0.0001	0.003	0.0007
Illumination angle	S	B	0.0004	0.005	0.006
Spectral product	S	B	0.001	0.0006	0.0003

Table 3.4b. Components of uncertainty which are independent of the scattering properties of the materials, and the resulting relative standard uncertainties

Component of uncertainty	Effect	Type	Relative standard uncertainty [%]		
			Standard geometry		
			20°	60°	85°
Primary gloss standard	S	B	0.1	0.06	0.005
Source stability	R	A	0.01		
DVM	S	B	0.002		
Signal noise	R	A	0.01		
Photodiode linearity	S	A	0.1		
Amplifier gain ratio	S	A	0.003		
Repeatability	R	A	0.1		

Table 3.5. Relative expanded uncertainties of specular gloss measured by the NIST reference goniophotometer ($k = 2$) for the 20°, 60°, and 85° standard geometries.

Standard geometry	Relative expanded uncertainty [%]
20°	0.35
60°	0.31
85°	0.29

4. Primary Gloss Standard

The selection criteria for the new NIST primary specular gloss standard are described, along with the index of refraction, specular gloss, and luminous reflectance properties of the standard. Finally, the new primary standard is compared to other gloss standards.

4.1 Description

The theoretical standard for specular gloss measurements is specified to be a highly polished plane black glass with an index of refraction at the wavelength of the sodium D line, 589.3 nm, of $n_D = 1.567$ and is assigned a specular gloss value of 100 for each of the three standard geometries. The specular reflectances of unpolarized light for this index of refraction are calculated using the Fresnel equations and are listed in Table 2.2. Since there is no black glass with exactly $n_D = 1.567$, other materials are used as primary standards and their gloss values are computed from the Fresnel reflectances. The primary specular gloss standard previously used at NIST in the 1970's was a commercial Carrara black glass with $n_D = 1.527$ [6]. A recent evaluation of this type of glass and its constituents was performed, and no match to any commercially available optical glass could be determined.

In general, there are a number of disadvantages to using black glass as a primary specular gloss standard. The surface of black glass is inhomogeneous, ages poorly, and is easily damaged, and thus requires frequent repolishing and recalibration. Polished black glass has been reported to be unsTable due to surface chemical contamination, resulting in variations of 0.3 % to 0.5 % in the index of refraction over a period of three to four years [19]. These variations correspond to changes of about 2 % and 1 % in the specular gloss values for the 20° and 60° standard geometries, respectively. Optical polishing with cerium oxide can restore the original gloss value. The uniformity of the index of refraction has been investigated [20], and varies across the surface by approximately 0.5 %.

The ISO 2813 documentary standard suggests an alternative to black glass – a clear glass with roughened edges and back surface, with the back surface painted black to absorb any transmitted light. A difficulty with such a primary standard is finding a black paint that provides a good match to the index of refraction of the clear glass. Otherwise, some scattered light from the back surface will enter the receiver and cause an error in the measured luminous flux. An alternative to the black paint is to cut the clear glass at an angle so that back reflections are not incident on the receiver, but this option is not feasible for small, porTable instruments commonly used in industry.

The disadvantages of the previous primary standards for specular gloss described above motivated a search for a new primary standard at NIST [21]. The following selection criteria were used: First, the standard should be a commercially available, high-purity optical glass with high chemical and mechanical durability. Second, the index of refraction n_D should be as close to 1.567 as possible to conform to the documentary standards. Third, the material should be homogenous and have dispersion characteristics similar to black glass. An optical quality barium crown glass BaK50 was selected as the new NIST primary standard since it possesses high chemical and mechanical durability and $n_D = 1.5677$. In addition, the index of refraction homogeneity of this glass is within 5×10^{-6} over an area of 70 mm^2, which is better than the black glass. Three pieces of BaK50 with dimensions of 98 mm by 98 mm by 20 mm were purchased, and the NIST Optical Shop fabricated sa0mples with a 6° wedge at the back surface. This angle is sufficient to reflect the light incident on the back surface out of the field of view of the receiver for all of the standard geometries, as shown in Fig. 4.1. The front and back surfaces were polished to a roughness of 0.6 nm to 1.0 nm. A black felt material was placed at the back and edges of the samples to eliminate reflections from the surrounding black anodized holder.

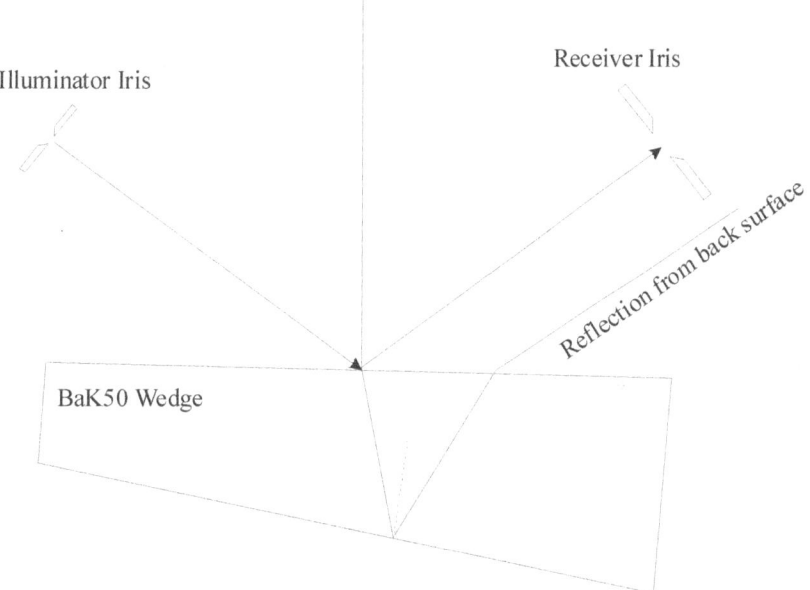

Figure 4.1. Schematic diagram of the new NIST primary standard showing the incoming and reflected beam at aspecular geometry (not to scale)

4.2 Characterization

The specular gloss value of the primary standard is determined relative to the theoretical gloss standard from the ratio of the luminous reflectance to the Fresnel reflectance of the theoretical standard, as shown in eq (2.10). The luminous flux reflected from the primary standard is used to determine the specular gloss of the test sample, from eq (2.8). There are two methods for determining the luminous reflectance of the primary standard, one from measurements of the index of refraction over the visible wavelength range, the other from direct measurements of this reflectance.

4.2.1 Index of Refraction The index of refraction of BaK50 was measured on a prism-shaped sample prepared from the same melt as the wedge samples. The minimum deviation technique [22] was used to determine the index of refraction at five different wavelengths, including 589.3 nm, with a standard uncertainty of 4×10^{-5} ($k = 2$). The indexes of refraction as a function of wavelength are given in Table 4.1. The data was fitted with a cubic spline over the range of data and linearly extrapolated to give the dispersion curve for wavelengths from 380 nm to 780 nm at 10 nm intervals, shown in Fig. 4.2. The calculated specular reflectances $\rho_s(\theta, \lambda_D)$ and specular gloss value $G_s(\theta)$ of the new NIST primary standard at $\lambda_D = 589.3$ nm are listed in Table 4.2 for the standard geometries.

Table 4.1. Index of refraction n of BaK50 glass as a function of wavelength

Wavelength [nm]	n
435.8	1.5800
480.0	1.5753
546.1	1.5702
589.3	1.5677
643.9	1.5654

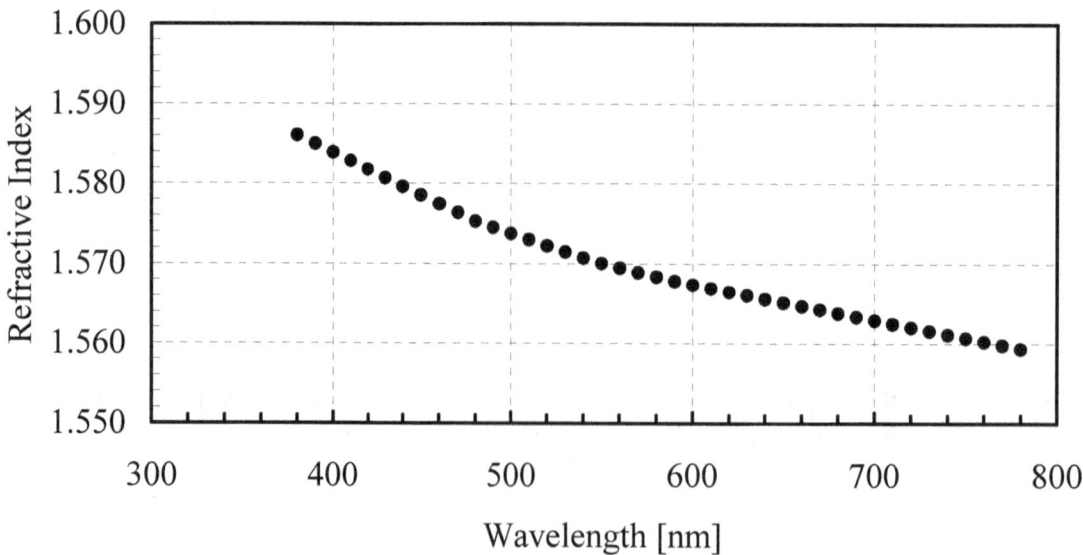

Figure 4.2. Fitted refractive index, $n(\lambda)$, for BaK50 glass as a function of wavelength

Table 4.2. Calculated specular reflectance $\rho_s(\theta,\lambda_D)$ and specular gloss value $G_s(\theta)$ of the new primary standard BaK50 at $\lambda_D = 589.3$ nm for each standard geometry

Standard geometry	$\rho_s(\theta, \lambda_D)$	$G_s(\theta)$
20°	0.049172	100.19
60°	0.100167	100.11
85°	0.619204	100.01

4.2.2 Luminous Reflectance There are two methods to determine the luminous reflectance of the primary standard. The first calculates the luminous reflectance from the index of refraction as a function of wavelength, while the second measures the luminous reflectance directly. The luminous reflectances and specular gloss values from these two methods are listed in Table 4.3, with the details of the calculations and measurements given below.

Table 4.3. Average luminous reflectance $\rho_{v,s}(\theta)$ and specular gloss value $G_s(\theta)$ for the new primary standard BaK50 for each standard geometry and method of calculation or measurement

	Standard geometry					
	20°		60°		85°	
Method	$\rho_{v,s}(\theta)$	$G_s(\theta)$	$\rho_{v,s}(\theta)$	$G_s(\theta)$	$\rho_{v,s}(\theta)$	$G_s(\theta)$
Index of refraction	0.049464	100.8	0.100510	100.5	0.619375	100.0
Luminous reflectance	0.049587	101.0	0.100743	100.7		

For the first method, the index of refraction as a function of wavelength, from Table 4.1 and Fig. 4.2, is used in eqs (2.11) to (2.13) to calculate $\rho_s(\theta, \lambda)$ for each of the standard geometries from 380 nm to 780 nm every 10 nm. The luminous reflectance $\rho_{v,s}(\theta)$ is calculated from Eq. (2.14) using $\rho_s(\theta, \lambda)$, CIE standard illuminant C for $S(\lambda)$, and the CIE spectral luminous efficiency function. Finally, the specular gloss value $G_s(\theta)$ for the primary standard is given by eq (2.10) but with $\rho_{v,s}(\theta)$ in place of ρ_s.

The second method measures luminous reflectance by an absolute technique using the NIST reference goniophotometer. For a fixed specular geometry, the following steps are followed for absolute luminous reflectance measurements. The sample is manually removed from the path of the influx beam, the receiver is rotated into the beam path, and the net illumination signal is measured. The sample is then placed in the beam path, the sample Table and receiver arm are rotated to the desired geometry, and the net reflected signal is measured. The net illumination signal is measured again, and the two values are averaged to obtain the final illumination signal. The luminous reflectance is calculated from the ratio of the net reflected signal to the net illumination signal. From the measured luminous reflectance of the primary standard, the specular gloss value was calculated as described in the previous paragraph. The 85° geometry was not measured due to limitations in the instrument. The relative expanded uncertainty ($k=2$) for the luminous reflectance of BaK50 at the specular geometries of 20° and 60° is 0.22 %. The random effects, which include the source stability and detector noise, result in a relative expanded uncertainty ($k=2$) of 0.1 % for both geometries. The systematic effects, which include the DVM accuracy, amplifier gain, detector linearity, source polarization, angular scale and spectral product, result in a relative expanded uncertainty ($k=2$) of 0.2 % for both geometries.

4.3 Comparison of Specular Gloss Standards

The specular gloss values of the new primary standard are nearly 100 for all three standard geometries as a consequence of n_D being close to the theoretical value of 1.567. Comparing Tables 4.2 and 4.3, the specular gloss values calculated from the luminous reflectance are greater than those calculated using only n_D for the 20° and 60° standard geometries, and are approximately equal for the 85° standard geometry. Finally, the luminous reflectances and corresponding specular gloss values obtained from each of the two methods, from Table 4.3, agree well with each other, the maximum difference in $G_s(\theta)$ being only 0.2. The uncertainty analysis shows that the calculated gloss value following the two different luminous reflectance procedures agree within the uncertainties.

The dispersion characteristics of BaK50 were compared to those of black glass and quartz gloss standards. Three different types of highly polished black glasses were investigated - the gloss standard previously used at NIST and two currently used in industry. Neither black glass nor quartz are a good match for $n_D = 1.567$. The refractive indexes for the black glass samples were measured using an Abbe refractometer at three different wavelengths and fit with a cubic spline over the range of data and linearly extrapolated to give the dispersion curves, from 380 nm to 780 nm at 10 nm intervals, shown in Fig. 4.3. The uncertainty of these measurements is 0.05 % ($k=2$). The refractive indexes for quartz listed in Ref. [23] was fitted using a cubic spline function over the range of the data and linearly extrapolated to give the dispersion curve for wavelengths from 380 nm to 780 nm at 10 nm interval shown in Fig. 4.3. The normalized dispersion curves of BaK50, black glass, and quartz standards are shown in Fig. 4.4. The index of refraction is normalized at a wavelength of 560 nm and plotted as a function of wavelength. The average of the dispersion curves for the black glass samples is plotted. The relative dispersion characteristic of BaK50 closely resembles that of the black glass samples, while quartz has a different characteristic. The international and national standards define the specular gloss value for the theoretical and working standards based upon a single refractive index, n_D, but the instruments are specified for luminous reflectance measurements. This ambiguity leads to a situation where the gloss value of the sample under test depends on the dispersive characteristics of the secondary-working standard, such as black glass. This is particularly important for instruments whose spectral characteristics are in poor agreement with those specified by the documentary standards. Ideally, the calibration of a test sample should not be affected by the properties of the working standard. This ambiguity produces differences in practical gloss measurements of up to 0.5 % since this recommendation ignores the dispersion characteristic of the material represented by changes in refractive index as a function of wavelength.

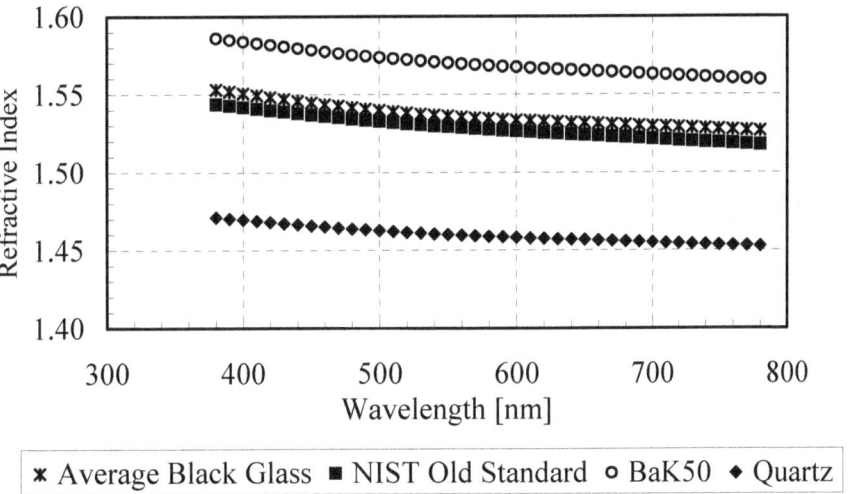

Figure 4.3. Fitted refractive index, $n(\lambda)$, for average black glass working standards, NIST old standard (Carrara black glass), BaK50, and quartz as a function of wavelength

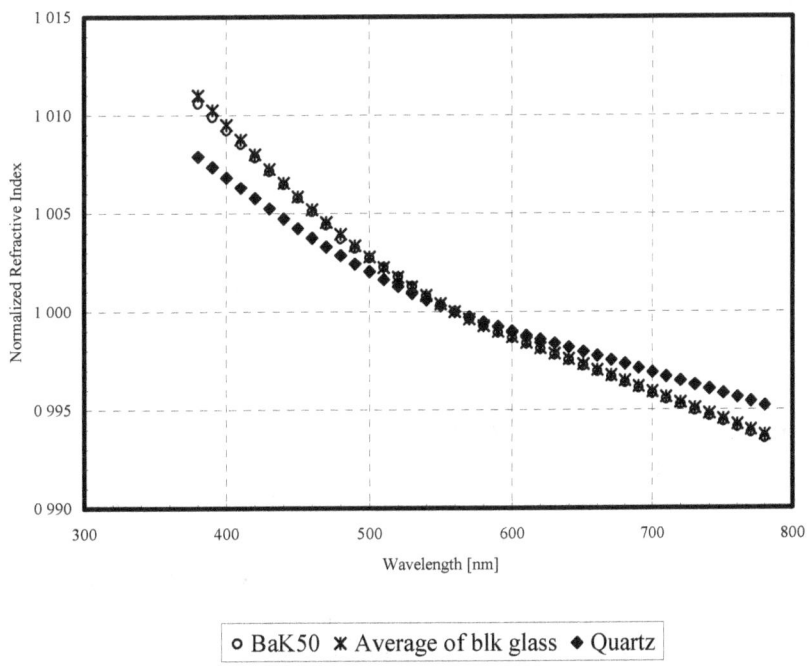

Figure 4.4. Normalized refractive index for BaK50, average of black glass working standards, and quartz as a function of wavelength

Acknowledgments

The success of the measurement service for specular gloss was aided by the efforts of Albert Parr and Sally S. Bruce.

References

[1] NIST Calibration Services Users Guide, NIST Special Publication 250.
[2] R. S. Hunter and R. W. Harold, The Measurement of Appearance, 2nd Ed., John Wiley and Sons, Inc., New York (1987).
[3] International Standard ISO 2813, Paint and Varnishes-Measurements of Specular Gloss of Nonmetallic Paint Films at 20°, 60°, and 85° (International Organization for Standardization 1978).
[4] Standard Test Method for Specular Gloss, ASTM D523, American Society for Testing and Materials, West Conshohocken, PA (1995).
[5] Commission International de l'Eclairage: International Lighting Vocabulary, Colorimetry, 2nd ed., Publ. No. 15.2 (1986).
[6] J. J. Hsia, The NBS 20-, 60-, and 85- Degree Specular Gloss Scales, NIST Tech. Note **594-10** (1975).
[7] F. E. Nicodemus, J. C. Richmond, J. J. Hsia, I. W. Ginsburg, and T. Limperis, Geometrical Considerations and Nomenclature for Reflectance, U.S. NBS Monograph **160** (1977).
[8] W. Budde, The Calibration of Gloss Reference Standards, Metrologia **16**, 89-93 (1980).
[9] M. E. Nadal and E. A. Thompson, NIST Reference Goniophotometer for Geometrical Appearance Measurements, Journal Coatings Technology, **73**, 73-80 (2001).
[10] R. S. Hunter, Gloss Investigations Using Reflected Images of a Target Patters, Journal of Research, **16**, 359 (1936).
[11] R. S. Hunter, Methods of Determining Gloss, NBS Research paper RP 958, Journal of Research, **18**, No. 77, 281 (1937).
[12] I. Nimeroff, Analysis of Goniophotometric Reflective Curves, Journal of Research, **48**, No. 5, 441 (1952).
[13] Y. Ohno, Photometric Calibrations, NIST SP **250-37** (1997).
[14] P. Y. Barnes and J. J. Hsia, 45°/0° Reflectance Factor of Pressed Polytetrafluoroethylene (PTFE) Powder, NIST Technical Note **1413** (1995).
[15] T. Doiron and J. Stoup, Uncertainty and Dimensional Calibrations, J. Res. Natl. Inst. Stand. Technol. **102**, 647 (1997).
[16] R. D. Saunders and J. B. Shumaker, Automated Radiometric Linearity Tester, Appl. Opt. **23**, 3504 (1984).
[17] T. C. Larson, S. S. Bruce, and A. C. Parr, Spectroradiometric Detector Measurements, NIST SP **250-41** (1998).
[18] B. N. Taylor and C. E. Kuyatt, Guidelines for Evaluating and Expressing the Uncertainty of NIST Measurements Results, NIST Tech. Note **1297** (1994).
[19] J. Zwinkels and M. Nöel, Specular Gloss Measurement Services at the National Research Council of Canada, Surface Coatings International **12**, 512 (1995).
[20] B. Wolfang and C. X. Dodd, Stability Problems in Gloss Measurements, Journal of Coating Tech. **552**, 44-48, (1980).
[21] M. E. Nadal and E. A. Thompson, New Primary Standard for Specular Gloss Measurements, Journal of Coatings Technology, **72**, No. 911 (2000).
[22] G. E. Fishter, Refractometry, Applied Optics and Optical Engineering, Vol. IV, Academic Press, New York (1967) p. 363-382.
[23] I. H. Malitson, Interspecimen Comparison of the Refractive Index of Fused Silica, Journal of the Optical Society of America **55,** 1205-1209 (1965).

Appendix A

REPORT OF CALIBRATION

38090S Specular Gloss

for

Specular Gloss Plates

Submitted by:

Any Company, Inc.
Attn.: Ms. Jane Doe
123 Calibration Street
Measurement City, MD 20800-1234

(See your Purchase Order No. 12345, dated month Day, Year)

1. Description of Calibration Items

Twelve specular gloss plates, 10.8 cm square, manufactured by Hunter Associates Laboratory, Inc., with serial numbers SG-01, SG-02, SG-03, SG-04, SG-05, SG-06, SG-07, SG-08, SG-09, SG-10, SG-11, and SG-12, submitted by Any Company, Inc.

2. Description of Calibration

Specular gloss is a measure of the specular luminous reflectance of an item under geometrical and spectral conditions specified by the American Society for Testing Materials (ASTM) D523 [1] and the International Organization for Standardization (ISO) 2813 [2] documentary standards.

The calibration items were measured using the NIST Reference Goniophotometer for Specular Gloss. This instrument was originally developed at NIST [3] in 1970 and recently updated, automated, and characterized [4] to comply with the requirements in the documentary standards. An achromatic condenser lens system images radiant flux from a quartz-tungsten-halogen incandescent lamp onto the source aperture. The beam emerging from the aperture is collimated by a lens, passes through a polarization scrambler, a color filter, and an iris with a clear diameter of 3 cm, and is incident upon the sample holder. The radiant flux reflected by the item in the sample holder is collected by the receiver iris and focused by a lens onto the receiver aperture at the port of an integrating sphere. The integrating sphere has a diameter of 30.5 cm and is packed

REPORT OF CALIBRATION
38090S Specular Gloss
Any Company, Inc.

Specular Gloss Plates
Serial Nos.: SG-01, SG-02, SG-03, SG-04,
SG-05, SG-06, SG-07, SG-08,
SG-09, SG-10, SG-11, SG-12

with pressed polytetrafluoroethylene (PTFE). A Si photodiode with a photopic filter is attached to the sphere and produces a signal proportional to the radiant flux in the sphere, and hence to the radiant flux reflected by the item in the sample holder. The sample holder and receiver are located on independent, aligned rotation stages used to select the angles of illumination and viewing.

Measuring the specular gloss of a calibration item is a relative measurement and therefore requires comparison to a standard with a calibrated specular gloss. The standard is the NIST primary specular gloss standard [5], a wedged piece of BaK50 glass.

The calibration items and primary gloss standard were cleaned with an air bulb and sequentially mounted in the sample holder so that their front surface was aligned with the axis of rotation of the rotation stage and the illumination beam was centered on the front surface. The calibration items were aligned in the direction of illumination, as indicated on the back of the sample. Measurements of the calibration items were bracketed with measurements of the primary specular gloss standard to correct for any instrumental drift.

The documentary standards specify the geometrical conditions of measurement for three geometries, designated by illumination angles of 20°, 60°, and 85°. For all three geometries, the angle from the center of the collimating lens of the illuminator to the source aperture was 0.75° in the plane of illumination and 3° perpendicular to this plane. For the 20° geometry, the angle from the center of the converging lens of the receiver to the receiver aperture was 1.8° in the plane of illumination and 3.6° perpendicular to this plane. For the 60° and 85° geometries, these angles were 4.4° and 11.7° and 4° and 6°, respectively. The documentary standards also specify the spectral conditions. The spectral power distribution of the illuminator was that of CIE standard illuminant C and the spectral responsivity was that of the CIE spectral luminous efficiency function V_λ. Both the geometrical and spectral conditions of the NIST Reference Goniophotometer for Specular Gloss fell within the tolerances given in the documentary standards.

3. Results of Calibration

The specular gloss G is given by

$$G(\theta_0) = 100 \cdot \frac{\rho_{s,v}(\theta_0)}{\rho_0(\theta_0, \lambda_D)} \cdot \frac{I}{I_s} , \qquad (3.1)$$

where θ_0 is the angle of illumination (20°, 60°, or 85°), λ_D is the wavelength of the sodium D-line (589.3 nm), ρ_s is the reflectance of the NIST primary gloss standard, ρ_0 is the reflectance of the theoretical standard ($n_D = 1.567$), and I and I_s are the currents from the receiver when measuring the calibration item and primary standard, respectively.

REPORT OF CALIBRATION
38090S Specular Gloss
Any Company, Inc.

Specular Gloss Plates
Serial Nos.: SG-01, SG-02, SG-03, SG-04,
SG-05, SG-06, SG-07, SG-08,
SG-09, SG-10, SG-11, SG-12

The final specular gloss is obtained by averaging the values from multiple measurements. The certified 20°, 60°, and 85° specular gloss values for the calibration items are given in Tables 1 to 3, respectively.

Uncertainties were calculated according to the procedures outlined in [6]. Sources of uncertainty due to random effects are source stability, detector noise, and item uniformity. The uncertainty contribution caused by these effects was evaluated from the standard deviation of repeat measurements of each item.

Sources of uncertainty due to systematic effects included those that depend on the scattering properties of the calibration item (polarization, divergence and spectral power distribution of the illuminator, illumination angle, and spectral responsivity of the receiver) and those that are independent of the scattering properties (detector linearity, amplifier gain ratio, and voltmeter accuracy). The uncertainties from these effects were evaluated as part of the characterization of the NIST Reference Goniophotometer for Specular Gloss, taking into account the spectral reflectance properties of the calibration items. All uncertainty components were assumed to have normal probability distributions.

The resulting uncertainty contributions to specular gloss values due to systematic and random effects are also given in Tables 1 to 3. The expanded uncertainty was obtained from the root-sum-square of the uncertainty contributions multiplied by a coverage factor $k = 2$. The expanded uncertainties in specular gloss are also given in Tables 1 to 3.

4. General Information

1) The calibration items were measured in the "as received" condition after cleaning using an air bulb.

2) This calibration report may not be reproduced except in full without the written consent of this Laboratory.

Prepared by:

Approved by:

Maria E. Nadal
Optical Technology Division
Physics Laboratory
(301) 975-4632

Gerald T. Fraser
For the Director,
National Institute of
Standards and Technology
(301) 975-3797

Calibration Date: Month Day, year
NIST Test No.: Division number/Test Folder Number-Year

REPORT OF CALIBRATION
38090S Specular Gloss Specular Gloss Plates
Any Company, Inc. Serial Nos.: SG-01, SG-02, SG-03, SG-04,
  SG-05, SG-06, SG-07, SG-08,
  SG-09, SG-10, SG-11, SG-12

References:

[1] "Standard Test Method for Specular Gloss, ASTM D523," American Society for Testing and Materials, West Conshohocken, PA (1995).

[2] International Standard ISO 2813, "Paint and Varnishes-Measurements of Specular Gloss of Nonmetallic Paint Films at 20°, 60°, and 85°" (International Organization for Standardization 1978).

[3] Hsia, J. J., "The NBS 20-, 60-, and 85- Degree Specular Gloss Scales," NIST Tech. Note 594-10 (1975).

[4] Nadal, M. E. and Thompson, E. A., "NIST Reference Goniophotometer for Reflection Gloss Measurements," Journal Coatings Technology, vol. 73, 73-80, June 2001.

[5] Nadal, M. E. and Thompson, E. A., "New Primary Standard for Specular Gloss Measurements," Journal of Coatings Technology, Vol. 72, December 2000.

[6] B. N. Taylor and C. E. Kuyatt, "Guidelines for Evaluating and Expressing the Uncertainty of NIST Measurement Results," NIST Technical Note 1297 (1994).

Calibration Date: Month Day, year
NIST Test No.: Division number/Test Folder Number-Year

REPORT OF CALIBRATION
38090S Specular Gloss
Any Company, Inc.

Specular Gloss Plates
Serial Nos.: SG-01, SG-02, SG-03, SG-04,
SG-05, SG-06, SG-07, SG-08,
SG-09, SG-10, SG-11, SG-12

Table 1. 20° specular gloss, uncertainty contributions, and expanded uncertainties ($k = 2$) for the calibration items.

Item Serial Number	Specular Gloss	Uncertainty Contribution		Expanded Uncertainty ($k = 2$)
		Random	Systematic	
SG-01	9.00	0.35	0.13	0.75
SG-02	12.00	0.18	0.13	0.44
SG-03	42.12	0.13	0.13	0.37
SG-04	64.35	0.29	0.13	0.64

Table 2. 60° specular gloss, uncertainty contributions, and expanded uncertainties ($k = 2$) for the calibration items.

Item Serial Number	Specular Gloss	Uncertainty Contribution		Expanded Uncertainty ($k = 2$)
		Random	Systematic	
SG-05	5.87	0.35	0.11	0.73
SG-06	13.33	0.37	0.11	0.77
SG-07	38.65	0.11	0.11	0.31
SG-08	90.41	0.03	0.11	0.23

Table 3. 85° specular gloss, uncertainty contributions, and expanded uncertainties ($k = 2$) for the calibration items.

Item Serial Number	Specular Gloss	Uncertainty Contribution		Expanded Uncertainty ($k = 2$)
		Random	Systematic	
SG-09	8.47	0.55	0.10	1.12
SG-10	43.26	0.48	0.10	0.98
SG-11	58.32	0.25	0.10	0.54
SG-12	85.41	0.30	0.10	0.63

Calibration Date: Month Day, year
NIST Test No.: Division number/Test Folder Number-Year

Appendix B

NIST Reference Goniophotometer for Specular Gloss Measurements

María E. Nadal and E. Ambler Thompson
Optical Technology Division
National Institute of Standard and Technology
Gaithersburg, MD 20899-8442

ABSTRACT

A new reference goniophotometer for specular gloss measurements has been assembled and characterized, and will provide calibrations with an expanded uncertainty ($k = 2$) of 0.3 gloss units.

Keywords: appearance attributes; gloss; gonio-apparent; goniophotometer; specular gloss; specular reflectance.

I. Introduction

The current global economy has increased international competition and the need to improve the quality of many manufactured products. Appearance measurements are used to determine the quality and acceptability of a variety of products from textiles to machine finishes. Since the manufacture and marketing of these products is international in scope, the methods, instruments, and standards for the measurement of appearance must be precise, reproducible, and internationally standardized as well.

The appearance of an object is the result of a complex interaction of the light incident on the object, the optical characteristics of the object, and human perception. The scientific basis for the measurement of appearance attributes has been under investigation for the better part of a century and was standardized by the Commission Internationale de l'Eclairage (CIE) [1]. The visual response of human observers to wavelengths of light for the spectral region of 360 nm to 830 nm is tabulated as the CIE standard observer. The CIE has defined a series of standard spectral distributions that are known as the CIE Standard Illuminants.

Appearance attributes of an object are roughly divided into chromatic and geometric. The chromatic attributes are those that vary the spectral distribution of the reflected light (color). The geometric attributes are those that vary the spatial or angular distribution of the reflected light (gloss and haze).

This paper describes the reference goniophotometer at the National Institute of Standard and Technology (NIST). This instrument complies with the geometrical and spectral conditions specified by the American Society for Testing Materials (ASTM) D523 [2] and the International Organization for Standardization (ISO) 2813 [3] documentary standards for the measurement of specular gloss of non-metallic surfaces at fixed specular geometries of 20°, 60°, and 85°. Gloss measurements for other industrial applications using different geometries can also be accommodated. This goniophotometer is also capable of characterizing the distribution of unabsorbed incident radiation by reflection or transmission from a material at incident angels from 0° to 85° in compliance with ASTM E167 [4]. Figure 1 shows a plot of the luminous reflectance as a function of the observation angle for a set of black glasses with different values of gloss.

This goniophotometer was originally developed at NIST [5] in the 1970's by scientists active in the development of the ASTM standard test method, and has recently been updated, automated, and characterized. The instrument was revived to meet industrial needs, such as the paint and automotive industries, for accurate gloss measurements and direct traceability to national standards.

The outline of this paper is as follows. Section II presents the gloss definition and the measurement equations. The instrument design is presented in Sec. III and the characterization of this instrument is described in Sec. IV. The uncertainty analysis is discussed in Sec. V. Finally, the conclusions are detailed in Sec. VI.

II. Gloss Definition and Measurement Equations

This section details the basic definitions and the relevant measurement equations to determine specular gloss. The approach utilized in this paper is based upon the concepts presented in Refs. [5-7].

Gloss is the perception by an observer of the shiny appearance of a surface. This perception changes whenever there is a change in the relative position or spectral distribution of the source, the sample, and the observer. Different standard geometries are used to determine the specular gloss of materials. Table 1 lists examples of these geometries and their applications.

These geometries were selected based on their ability to produce optimum discrimination between samples and to correlate with visual rankings.

The measurement of specular gloss compares the specular luminous reflectance from a test specimen to that from a standard surface under the same experimental conditions. The incident beam is a collimated white light source that simulates CIE standard Illuminant C and the amount of radiation in the reflected beam is measured with a photopic receptor. The theoretical standard for specular gloss measurements is a highly polished plane black glass with a refractive index for the sodium D line $n_D = 1.567$. To set the specular gloss scale, the specular gloss of this theoretical standard has an assigned value of 100 for each of the three standard specular geometries of 20°, 60°, and 85°. For primary standards with a refractive index different from $n_D = 1.567$, the specular gloss is computed from the refractive index and the Fresnel equations. Recommendations for specular gloss standards and a new NIST primary gloss standard are described in Ref. [8].

Measuring the specular gloss of a test sample involves the ratio of the luminous reflectances of the test sample and primary standard. The specular gloss of a test sample at a nominal angle θ_0 is given by

$$G_t(\theta_0) = G_s(\theta_0) \cdot \frac{\rho_{v,t}(\theta_0)}{\rho_{v,s}(\theta_0)}, \quad (1)$$

where $G_s(\theta_0)$ is the specular gloss of the primary standard and $\rho_{v,t}(\theta_0)$ and $\rho_{v,s}(\theta_0)$ are the specular luminous reflectances of the test sample and primary standard, respectively. Furthermore, the specular gloss of the primary standard is given by

$$G_s(\theta_0) = G_0(\theta_0) \cdot \frac{\rho_s(\theta_0, \lambda_D)}{\rho_0(\theta_0, \lambda_D)}, \quad (2)$$

where $G_0(\theta_0)$ is the specular gloss of the theoretical standard and $\rho_s(\theta_0, \lambda_D)$ and $\rho_0(\theta_0, \lambda_D)$ are the specular reflectances of the primary and theoretical standards, respectively, at a wavelength $\lambda_D = 589.3$ nm. Combining Eqs. (1) and (2), the specular gloss of the sample under test is given by

$$G_t(\theta_0) = G_0(\theta_0) \cdot \frac{\rho_s(\theta_0, \lambda_D)}{\rho_0(\theta_0, \lambda_D)} \cdot \frac{\rho_{v,t}(\theta_0)}{\rho_{v,s}(\theta_0)}. \quad (3)$$

For each of the three standard geometries (20°, 60°, and 85°), the specular gloss of the theoretical standard is defined as $G_0(\theta_0) = 100$. The specular reflectance ρ from the surface of a nonmetallic sample depends on the incident angle θ defined relative to the normal of the sample, the wavelength λ, and polarization σ (p or s) of the incident radiation. The specular reflectance as a function of these variables is given by the Fresnel equations,

$$\rho(\theta, \lambda, \text{p}) = \left[\frac{n^2(\lambda) \cdot \cos\theta - \sqrt{n^2(\lambda) - \sin^2\theta}}{n^2(\lambda) \cdot \cos\theta + \sqrt{n^2(\lambda) - \sin^2\theta}} \right]^2 \text{ and} \quad (4)$$

$$\rho(\theta, \lambda, \text{s}) = \left[\frac{\cos\theta - \sqrt{n^2(\lambda) - \sin^2\theta}}{\cos\theta + \sqrt{n^2(\lambda) - \sin^2\theta}} \right]^2 \quad (5)$$

The specular reflectance for unpolarized light is calculated from the average of Eqs. (4) and (5). The specular reflectances for unpolarized light for the theoretical standard $\rho_0(\theta_0, \lambda_D)$ with $n_D = 1.567$ are calculated from Eqs. (4) and (5) and are listed in Table 2. For the case of the new primary gloss standard [8], the specular reflectances are calculated from the measured $n_D = 1.5677 \pm 0.0004$ and Eqs. (4) and (5).

The experimentally measured quantity is the ratio of the luminous fluxes reflected from the test sample and primary standard, $\Phi_{v,t,r}$ and $\Phi_{v,s,r}$ respectively,

$$\frac{\rho_{v,t}(\theta_0)}{\rho_{v,s}(\theta_0)} = \frac{\Phi_{v,s,i}}{\Phi_{v,t,i}} \cdot \frac{\Phi_{v,t,r}}{\Phi_{v,s,r}} = \frac{\Phi_{v,s,i}}{\Phi_{v,t,i}} \cdot \frac{R_s}{R_t} \cdot \frac{N_t}{N_s} \cdot \frac{A_s}{A_t} \tag{6}$$

where R is the responsivity of the receptor, N is the measured signal, A is the amplifier gain factor, and the subscripts s and t denote standard and test sample, respectively. Therefore, combining Eqs. (3) and (6), the specular gloss of a test sample in terms of the measured quantities reduces to

$$G_t(\theta_0) = 100 \cdot \frac{\rho_s(\theta_0, \lambda_D)}{\rho_0(\theta_0, \lambda_D)} \cdot \frac{\Phi_{v,s,i}}{\Phi_{v,t,i}} \cdot \frac{R_s}{R_t} \cdot \frac{N_t}{N_s} \cdot \frac{A_s}{A_t}. \tag{7}$$

The stability of the incident flux and responsivity are determined from the characterization of the instrument.

In terms of spectral and geometrical quantities, the luminous reflectance ρ_v of a sample at a nominal angle θ_0 is given by

$$\rho_v(\theta_0) = \frac{\sum_\sigma \int d\theta \int d\lambda \cdot S(\theta,\lambda,\sigma) \cdot \rho(\theta,\lambda,\sigma) \cdot V(\theta,\lambda,\sigma)}{\sum_\sigma \int d\theta \int d\lambda \cdot S(\theta,\lambda,\sigma) \cdot V(\theta,\lambda,\sigma)}, \tag{8}$$

where S is the incident spectral flux distribution, V is the relative response function, and ρ is the reflectance of the sample. These variables depend on the angle of incidence θ, wavelength λ, and polarization σ of the incident beam. In practice, the luminous reflectances of the standard and test sample are not measured separately. Rather, the ratio of these reflectances is measured and is given by

$$\frac{\rho_{v,t}(\theta_0)}{\rho_{v,s}(\theta_0)} = \frac{\sum_\sigma \int d\theta \int d\lambda \cdot S(\theta,\lambda,\sigma) \cdot \rho_t(\theta,\lambda,\sigma) \cdot V(\theta,\lambda,\sigma)}{\sum_\sigma \int d\theta \int d\lambda \cdot S(\theta,\lambda,\sigma) \cdot \rho_s(\theta,\lambda,\sigma) \cdot V(\theta,\lambda,\sigma)} \tag{9}$$

Then, the specular gloss of a test sample is given by

$$G_t(\theta_0) = 100 \cdot \frac{\rho_s(\theta_0, \lambda_D)}{\rho_0(\theta_0, \lambda_D)} \cdot \frac{\sum_\sigma \int d\theta \int d\lambda \cdot S(\theta,\lambda,\sigma) \cdot \rho_t(\theta,\lambda,\sigma) \cdot V(\theta,\lambda,\sigma)}{\sum_\sigma \int d\theta \int d\lambda \cdot S(\theta,\lambda,\sigma) \cdot \rho_s(\theta,\lambda,\sigma) \cdot V(\theta,\lambda,\sigma)} \tag{10}$$

Following the standard recommendations, the incident spectral flux distribution $S(\lambda)$ is specified to be CIE standard Illuminant C $S_C(\lambda)$ and the relative response function $V(\lambda)$ is specified to be the CIE 1931 2° observer spectral luminous efficiency function V_λ. In addition, the incident beam of light is specified to be collimated and unpolarized at a nominal incident angle θ_0. The specular gloss of the sample under test for the ideal case is given by

$$G_{t,id}(\theta_0) = 100 \cdot \frac{\rho_s(\theta_0, \lambda_D)}{\rho_0(\theta_0, \lambda_D)} \cdot \frac{\int d\lambda \cdot S_C(\lambda) \cdot \rho_t(\theta_0, \lambda) \cdot V_\lambda(\lambda)}{\int d\lambda \cdot S_C(\lambda) \cdot \rho_s(\theta_0, \lambda) \cdot V_\lambda(\lambda)}. \tag{11}$$

Deviations of the reference instrument from the ideal case include deviations from the specified spectral flux distribution of the source, spectral responsivity of the receptor, unpolarized and perfectly collimated incident beam, and nominal incident angle. The correction factors for the specular gloss of the sample under test as determined with the reference instrument are obtained using Eqs. (10) and (11). The correction factors are the ratio of the specular gloss for the ideal case to the specular gloss for the reference instrument

$$C = \frac{G_{t,id}(\theta_0)}{G_t(\theta_0)}. \tag{12}$$

The correction factor for deviations from the ideal spectral flux distribution is given by

$$C_s = \frac{\int d\lambda \cdot S_C(\lambda) \cdot \rho_t(\theta_0, \lambda) \cdot V_\lambda(\lambda)}{\int d\lambda \cdot S_C(\lambda) \cdot \rho_s(\theta_0, \lambda) \cdot V_\lambda(\lambda)} \cdot \frac{\int d\lambda \cdot S(\lambda) \cdot \rho_s(\theta_0, \lambda) \cdot V(\lambda)}{\int d\lambda \cdot S(\lambda) \cdot \rho_t(\theta_0, \lambda) \cdot V(\lambda)}. \tag{13}$$

The correction factor for deviations from an unpolarized incident beam is given by

$$C_p = \frac{\int d\lambda \cdot S_C(\lambda) \cdot \rho_t(\theta_0, \lambda) \cdot V_\lambda(\lambda)}{\int d\lambda \cdot S_C(\lambda) \cdot \rho_s(\theta_0, \lambda) \cdot V_\lambda(\lambda)} \cdot \frac{\sum_\sigma \int d\lambda \cdot S_C(\lambda, \sigma) \cdot \rho_s(\theta_0, \lambda, \sigma) \cdot V_\lambda(\lambda)}{\sum_\sigma \int d\lambda \cdot S_C(\lambda, \sigma) \cdot \rho_t(\theta_0, \lambda, \sigma) \cdot V_\lambda(\lambda)}. \tag{14}$$

The correction factor for deviations from a perfectly collimated incident beam is given by

$$C_d = \frac{\int d\lambda \cdot S_C(\lambda) \cdot \rho_t(\theta_0, \lambda) \cdot V_\lambda(\lambda)}{\int d\lambda \cdot S_C(\lambda) \cdot \rho_s(\theta_0, \lambda) \cdot V_\lambda(\lambda)} \cdot \frac{\int d\theta \int d\lambda \cdot S_C(\theta, \lambda) \cdot \rho_s(\theta, \lambda) \cdot V_\lambda(\lambda)}{\int d\theta \int d\lambda \cdot S_C(\theta, \lambda) \cdot \rho_t(\theta, \lambda) \cdot V_\lambda(\lambda)}. \tag{15}$$

The correction factor for deviations from the nominal incident angle is given by

$$C_a = \frac{\int d\lambda \cdot S_C(\lambda) \cdot \rho_t(\theta_0, \lambda) \cdot V_\lambda(\lambda)}{\int d\lambda \cdot S_C(\lambda) \cdot \rho_s(\theta_0, \lambda) \cdot V_\lambda(\lambda)} \cdot \frac{\int d\lambda \cdot S_C(\lambda) \cdot \rho_t(\theta, \lambda) \cdot V_\lambda(\lambda)}{\int d\lambda \cdot S_C(\lambda) \cdot \rho_s(\theta, \lambda) \cdot V_\lambda(\lambda)}. \tag{16}$$

Therefore, the specular gloss for the ideal case in terms of the correction factors is given by

$$G_{t,id}(\theta_0) = G_t \cdot C_s \cdot C_p \cdot C_d \cdot C_a. \tag{17}$$

The final measurement equation, including the correction factors and the experimentally measured quantities, is given by

$$G_{t,id}(\theta_0) = 100 \cdot \frac{\rho_s(\theta_0, \lambda_D)}{\rho_0(\theta_0, \lambda_D)} \cdot \frac{\Phi_{v,s,i}}{\Phi_{v,t,i}} \cdot \frac{R_s}{R_t} \cdot \frac{N_t}{N_s} \cdot \frac{A_s}{A_t} \cdot C_s \cdot C_p \cdot C_d \cdot C_a. \tag{18}$$

III. Instrument Design

The reference goniophotometer is a monoplane gonio-instrument with a fixed source arm and a rotating sample table and receptor arm. The sample table and the receptor arm can be rotated independently of each other in the plane of incidence. A schematic diagram of the goniophotometer in reflectance mode is shown in Fig. 2 and a detailed description of the instrument is given below.

The source system consists of a lamp unit, a set of condenser lenses, a precision source aperture, a collimating lens, and a color filter. The light source is a quratz-tungsten-halogen lamp rated at 100 W. A constant dc current is run through the lamp from a computer controlled power supply. An image of the lamp filament is formed on the source aperture with the achromatic condenser lens system. This source aperture is the field stop for this optical system. The dimensions of the source aperture are within the tolerances of the documentary standard specifications as shown in Table 3. A second lens (focal length 192 mm) collimates the beam into an approximate 3 cm diameter beam at the sample holder. A color filter is used so that the spectral flux distribution of the source-filter combination closely resembles that of CIE Standard Illuminant C. A glass-laminated polarizer is used to provide linearly polarized light or a scrambler is used to provide unpolarized light in the incident beam.

The goniometer positions the sample and receptor for bi-directional reflectance and transmittance measurements. This goniometer consists of two rotation stages, which are computer controlled. The sample table is 480 mm in diameter and holds rectangular samples with dimensions up to 110 mm by 200 mm and thickness up to 35 mm. In addition, a sampler holder with vacuum suction is used for non-rigid samples such as paper. The optical components for the receptor arm are enclosed in a light-tight baffle tube with an iris diaphragm at the entrance. The collector lens (focal length 508 mm) is behind this iris diaphragm and focuses the reflected beam at the receptor aperture. Table 3 lists the recommended dimensions of the receptor aperture with specified tolerances and the measured dimensions for the NIST apertures. This aperture is located at the entrance of the sphere to restrict the field of view of the receptor. The receptor package consists of a 30.5 cm diameter sphere packed with pressed polytetrafluoroethylene (PTFE) and a photopic receptor. The photopic receptor is a temperature controlled silicon photodiode and a filter with a computer controlled variable gain current to voltage amplifier. The photodiode-filter combination approximates V_λ. The amplified output from the receptor is measured by a 6 1/2-digit voltmeter and sent to a computer.

Measurements of the sample under test are bracketed with primary specular gloss standard measurements in order to correct for any instrumental drift. The specular gloss of the test sample is then calculated from the ratio of the measured luminous flux reflected from the sample to the average of the measured luminous flux reflected from the primary gloss standard. If a polarizer is used, measurements are performed with parallel and perpendicular polarization with respect to the incident plane and the average of these two measurements yields the reflectance for unpolarized radiation.

IV. Characterization of Instrument

This instrument has been carefully characterized and calibrated to ensure proper operation and to verify agreement with the documentary standard specifications for specular gloss measurements - ASTM D523 and ISO 2813. These documentary standards specify the incident beam to be a collimated unpolarized beam, the specular angles, the source and receptor apertures, and the spectral flux distribution of the source and spectral response of the receptor. Details of the characterization of the instrument are given below.

The lamp is burned in at 100 W for 24 h and then at the operational current of 6.3 A for an additional 48 h prior to performing measurements. The stability of the source was investigated by measuring the flux incident on a Si photodiode at the sample location for a period of 30 min. A typical relative standard deviation over ten minutes is 0.01 %.

The degree of polarization of the source unit was measured by rotating a polarizer and measuring the maximum and minimum transmitted signals. The difference of the signals was divided by the sum to obtain the degree of polarization of the source. This ratio is approximately 9 %. Because the light source is not completely unpolarized, a glass-laminated polarizer or scrambler is used in the beam path just before the sample. The averaging sphere on the receptor arm in front of the photopic receptor randomizes any additional polarization imparted by the sample. The leakage of the glass-laminated polarizer for radiation polarized perpendicular to the intended direction of polarization on the source unit is 1.4 %. The degree of polarization of the source-scrambler combination is 0.02 %, providing a good beam of unpolarized incident radiation. The divergence of the source was determined experimentally by measuring the source size at specific distances. In the plane of measurement, the half-angle divergence of the source is 0.3°.

The recommended spectral condition for specular gloss measurements is for the source to simulate the spectral distribution of CIE standard Illuminant C and for the receptor response to simulate the CIE 1931 2° luminous efficiency function. The spectral distributions of these standards are shown in Fig. 3.

A color filter was designed so that the spectral flux distribution of the source-filter combination will closely resemble that of CIE Standard Illuminant C. The Correlated Color Temperature (CCT) and the spectral flux distribution of the source for the spectral range of 380 nm to 1070 nm were measured using a calibrated scanning spectroradiometer [9]. A freshly pressed polytetrafluoreethylene (PTFE) sample [10] was placed in the sample holder at an incident angle of 45° and an observation angle of ~0°. The spectral reflectance of PTFE is non-selective in the visible region. Therefore, the PTFE sample has no significant effect on the measured CCT of the source. The lamp current is set at 6.3 A to achieve a CCT of 2856 K ± 10 K with a coverage factor of $k = 2$. A color-temperature conversion filter was designed to simulate the spectral flux distribution of CIE standard Illuminant C (CCT = 6774 K), achieving a CCT of 6740 K ± 30 K with a coverage factor of $k = 2$.

The absolute spectral responsivity of the filter and receptor package was measured in the NIST Receptor Comparator Facility [11], and approximates that of the CIE 1931 2° standard observer. The measured spectral distribution was normalized to one at 560 nm to compare with the theoretical distribution. The responsivity spatial uniformity of the receptor was not determined since an integrating sphere is used in front of the receptor. Figure 4 shows a comparison of the spectral products of the source and receptor and of the CIE standard Illuminant C and the 1931 2° luminous efficiency function.

The angular scales of the sample table and receptor were checked for accuracy and repeatability from 0° to ± 85°. The incident angles on the sample table are calibrated using a set of isosceles prisms made by the NIST optical shop. These prisms have nominal base angles of 20°, 60°, and 85° with maximum deviations of 0.05°. These angles correspond to the standard geometries for evaluation of the specular gloss of nonmetallic paint samples. The prisms were calibrated by the Manufacturing Engineering Laboratory at NIST using a precision electronic autocollimator with an expanded uncertainty ($k = 2$) of 0.005° [12]. For the calibration of the incident angles, one of the calibrated prisms was mounted in the sample holder. The sample table was then rotated so that the back reflection was retro-reflected. This procedure was repeated for the three prisms, the resulting maximum uncertainty for the angular scale of the sample table is 0.05°. The repeatability of the angular scale for the sample table is 0.005°. Then, to calibrate the receptor angular scale, a front surface mirror replaced the prism and the reflection was centered in the receptor aperture by rotating the receptor arm. The maximum uncertainty for the receptor arm angular scale is 0.05° and the repeatability is 0.001°.

The dimensions of the source and receptor apertures are predetermined by the specular geometry, focal length of the collimator lens in the system, test material, and the correlation with visual ranking. Table 3 lists the standard recommended aperture sizes and acceptable tolerances and measured values for the NIST apertures. The linear dimensions of the apertures were measured and the angular dimensions were calculated. All of the apertures are within the ASTM tolerances.

The photopic receptor consists of a temperature controlled silicon photodiode with a photopic filter and a computer controlled variable gain current to voltage amplifier. The gain setting of the amplifier is the power of ten by which the current is multiplied to convert it to voltage. For low gloss samples, the measured signal is much lower than the signal from the high specular gloss primary standard. In this case, several amplifier gain settings are required to cover the dynamic range. From Eq. (7), the ratio of the gain is the required quantity. The gain settings of the amplifier were calibrated using a reference constant current source as the input of the amplifier and a voltmeter was used to measure the output signal. At each gain setting, the current was set to obtain a range of voltages from 10 V to 0.1 V in twenty equally spaced steps and the output signal was measured at each current. This protocol yielded current settings that were common to successive gain settings. The average value of the gain ratio and relative standard deviation of these overlapping settings for successive gains are listed in Table 4. The gain ratios are not exactly equal to 10, and for better results, it is best to measure signals between the ranges of 0.5 V to 12 V.

The linearity of the photodiode-amplifier combination was measured using the NIST automated beam conjoiner [13]. This facility uses the beam addition method with a set of filters in the path of the two branches of the beam to vary the radiant flux at the receptor by three decades. The linearity of the photodiode-amplifier combination was checked at gain settings from 5 to 9. The maximum radiant flux at the receptor is controlled by a set of neutral-density filters in front of the receptor. Figure 5 shows a plot of the relative responsivity (ratio of the measured signal to the actual flux) as a function of current for the gain settings listed in the legend. The measured relative responsivity of the photodiode-amplifier combination is linear within 0.2 % for gain settings of 5 to 8. Larger deviations are observed at gain 9 due to the small current at this setting.

V. Uncertainty Analysis

This section describes the components of uncertainty, their evaluation, and the resulting uncertainty for specular gloss measurements. The uncertainty analysis described below follows the guidelines given in Ref. [14]. The NIST reference goniophotometer was designed so that the various sources of uncertainty were minimized and the residual uncertainties were characterized to produce specular gloss with a minimum relative expanded uncertainty ($k = 2$) of 0.4 %.

The specular gloss of a test sample is not measured directly. In fact, specular gloss is determined from quantities such as incident and reflected fluxes through the measurement equation. In general, the value of a measurand, y, is obtained from n other quantities x_i through a functional relation, f, given by

$$y = f(x_1, x_2, \cdots, x_i, \cdots, x_n) \tag{19}$$

The standard uncertainty of an input quantity is the estimated standard deviation associated with this quantity and is denoted by $u(x_i)$. The standard uncertainties are classified by the effect of their source and the method of evaluation. The effect is either random or systematic. The method of evaluation is either Type A, which is based on statistical analysis, or Type B, which is based on other means. The relative standard uncertainty is given by $u(x_i)/x_i$. The relative combined standard uncertainty $u_c(y)$ is given by the root-sum-square of the standard uncertainties associated with each quantity x_i, assuming that these uncertainties are uncorrelated for multiplicative functional relationships. The mathematical expression for the relative combined standard uncertainty is

$$u_c^2(y)/y^2 = \sum_{i=1}^{n} \left(\frac{1}{y}\frac{\partial f}{\partial x_i}\right)^2 u^2(x_i), \tag{20}$$

where $(1/y)(\partial f/\partial x_i)$ is the relative sensitivity coefficient. The expanded uncertainty U is given by $k \cdot u_c(y)$, where k is the coverage factor and is chosen on the basis of the desired level of confidence to be associated with the interval defined by U. For the purpose of this paper, $k = 2$ will be used, which defines an interval with a level of confidence of 95 %.

The components of uncertainty associated with gloss measurements are divided into those arising from experimentally measured quantities and those from deviations from standard recommendations. The appropriate measurement equation for the first case is Eq. (7) and for the latter case Eqs. (13) to (17) are applicable. The resulting uncertainty from deviations from the standard recommendations depends on the nature of the sample under test and the primary standard. This uncertainty analysis was applied to representative measurements of a highly polished black glass as the test sample and to a highly polished wedge of BaK50 as the primary standard.

The component of uncertainty arising from the specular reflectance of the primary gloss standard is uncertainty in the refractive index measurements. The uncertainties from the refractive index measurements are a systematic effect with a Type B evaluation. The relative uncertainty resulting from the refractive index measurements is 0.1 %, 0.06 %, and 0.005 % for the 20°, 60°, and 85° geometries, respectively.

The components of uncertainty arising from the source are uncertainties from the stability, polarization, and divergence of the source. The source stability results in an uncertainty from a random effect with a Type A evaluation. The relative standard uncertainty due to instability of the source is 0.001 % for a measurement time of 10 min. The uncertainty arising from the polarization of the source is a systematic effect with a Type B evaluation and depends on the scattering properties of the material. The correction factor for deviations from an unpolarized incident beam were calculated from Eq. (14) using the degrees of polarization determined in the previous section. The correction factors for the three standard geometries are listed in Table 5. The uncertainty arising from divergence of the source is a systematic effect with a Type B evaluation and depends on the scattering properties of the material. The correction factor for deviations from a perfectly collimated incident beam were calculated from Eq. (15) using the divergence determined in the previous section. The correction factors for the three standard geometries are listed in Table 5.

The uncertainties arising from the goniometer are the uncertainties of the angular scales. This uncertainty is a systematic effect with a Type B evaluation and depends on the scattering properties of the sample. The angular setting uncertainty of 0.05° is smaller than the 0.1° recommended accuracy by the ASTM D523. The correction factors from deviations of the nominal incident angle from Eq. (16) for the three standard geometries are listed in Table 5.

The components of uncertainty arising from the detection system are uncertainties from the digital voltmeter (DVM), signal noise, amplifier gain ratio, and photodiode linearity. The uncertainty from the DVM is a systematic effect with a Type B evaluation, assuming a normal probability distribution. Using the manufacturer's specifications, the relative uncertainty resulting from the DVM is 0.002 %. Signal noise of the instrument results in an uncertainty from a random effect with a Type A evaluation. A typical relative standard deviation for a gloss measurement is 0.01 %. The uncertainties arising from the photodiode linearity and the amplifier gain ratio are systematic effects with a Type A evaluation. The maximum relative standard deviation from linearity of the photodiode, for signals in the range of 0.1 V to 12 V, is 0.09 %. The relative standard deviations for successive gain settings are listed in Table 4. The maximum relative standard deviation is 0.002 %.

The repeatability of the sample positioning results in an uncertainty from a random effect with a Type B evaluation. Assuming a rectangular probability distribution with a maximum deviation of 0.2 gloss units, the relative standard uncertainty is 0.1 %.

Deviations from the spectral condition are important for low-gloss materials with strong colors. The spectral conditions of the source and receptor are in close agreement with the recommended CIE Standard Illuminant C and CIE 1931 2° luminous efficiency function. The difference between the instrument and the standard recommendations results in a systematic effect with a Type B evaluation. The correction factors for deviations from the standard recommendations for the three standard geometries, from Eq. (13), are listed in Table 5.

Since all of the correction factors listed in Table 5 are nearly unity, no correction factor is applied and the variations from unity are considered to be uncertainties. The relative standard uncertainty from each component of uncertainty is listed in Tables 6a and 6b for sample-dependent and –independent components, respectively. The relative expanded uncertainties given in Table 7 are calculated from the root-sum-square of the uncertainties listed in Tables 6a and 6b. The ISO 2813 standard specifies that gloss should be reported to the nearest full gloss unit. Therefore, the calibration of the reference instrument and primary standard should have an uncertainty of less than 1 gloss unit. The NIST reference goniophotometer has an expanded ($k = 2$) relative uncertainty of 0.3 % for calibrating a highly polished black glass. If the specular gloss of successive measurements on the same sample do not agree to within 0.3 %, the instrument is assumed to have failed during the measurements and the sample is measured again.

VI. Conclusions

The appearance attributes of surfaces indicate the acceptability and quality of a large number of manufactured products. Specular gloss is the second most utilized attribute, after color to evaluate products such as automotive coatings, textiles, and papers. The reference goniophotometer described in this paper complies with the standard recommendations for specular gloss at 20°, 60°, and 85° geometries for non-metallic paint samples from low to high gloss levels, as described in the ISO 2813 and ASTM D523 documentary standards. In addition, this reference instrument is capable of performing bi-directional reflectance and transmission measurements at angles from 0° to 85° for both incident and viewing angles in compliance with the ASTM E 167 documentary standard. This instrument has been carefully characterized and calibrated to ensure proper operation and to verify agreement with the standard recommendations. The relative expanded uncertainty ($k = 2$) of the NIST goniophotometer for specular gloss is 0.3 % at all three standard geometries. The accuracy of specular gloss measurements depends not only on the properties of the instrument but also to a considerable extent on those of the primary gloss standard. A new primary gloss standard was developed at NIST and details on this standard characterization are the subject of another paper [8]. The new gloss standard and the NIST reference goniophotometer provide an accurate calibration facility for specular gloss.

Acknowledgements

The authors wish to express special thanks to Edward Early for many useful discussions and the collaboration with the National Institute of Standards and Technology researchers who are working on the Measurement Science for Optical Reflectance and Scattering Project.

References

[1] Commission International de l'Eclairage: International Lighting Vocabulary, Colorimetry, 2nd ed., Publ. No. 15.2 (1986).

[2] ASTM D523, "Standard Test Method for Specular Gloss," in ASTM standards on Color and Appearance Measurement, 5th ed., (ASTM, Philadelphia, PA, 1996).

[3] International Standard ISO 2813, "Paint and Varnishes-Measurements of Specular Gloss of Nonmetallic Paint Films at 20°, 60°, and 85°" (International Organization for Standardization 1978).

[4] ASTM E167, "Standard Practice for Goniophotometry of Objects and Materials," in ASTM standards on Color and Appearance Measurement, 5th ed., (ASTM, Philadelphia, PA, 1996).

[5] Hsia, J. J., "The NBS 20-, 60-, and 85- Degree Specular Gloss Scales," NIST Tech. Note 594-10 (1975).

[6] Nicodemus, F. E., Richmond, J. C., Hsia, J. J., Ginsburg, I. W., and Limperis, T., "Geometrical Considerations and Nomenclature for Reflectance," U.S. NBS Monograph 160 (1977).

[7] Budde, W., "The Calibration of Gloss Reference Standards," Metrologia 16, 89-93 (1980).

[8] Nadal, M. E. and Thompson, E. A., "New Primary Standard for Specular Gloss Measurements," submitted to JCT.

[9] Ohno, Y., "Photometric Calibrations," NIST SP 250-37 (1997).

[10] Barnes, P. Y. and Hsia, J. J., "45°/0° Reflectance Factor of Pressed Polytetrafluoroethylene (PTFE) Powder," NIST Technical Note 1413 (1995).

[11] Larson, T. C., Bruce, S. S., and Parr, A. C., "Spectroradiometric Receptor Measurements," NIST SP 250-41 (1998).

[12] Doiron, T. and Stoup, J., "Uncertainty and Dimensional Calibrations," J. Res. Natl. Inst. Stand. Technol. **102**, 647 (1997).

[13] Saunders, R. D. and Shumaker, J. B., "Automated Radiometric Linearity Tester," Appl. Opt. **23**, 3504 (1984).

[14] Taylor, B. N. and Kuyatt, C. E., "Guidelines for Evaluating and Expressing the Uncertainty of NIST Measurements Results," NIST Tech. Note 1297 (1994).

List of Figures

1. Luminous reflectance as a function of the observation angles for a set of black glasses with different levels of gloss an indicated in the legend.
2. A schematic diagram of the NIST reference goniophotometer.
3. Spectral flux distribution of the CIE Standard Illuminant C and CIE 1931 2° luminous efficiency function.
4. Actual and ideal spectral products of the source and receptor.
5. Relative responsivity (ratio of the measured signal to the actual flux) of the photodiode-amplifier combination as a function of current for the listed gain settings.

Figure 1

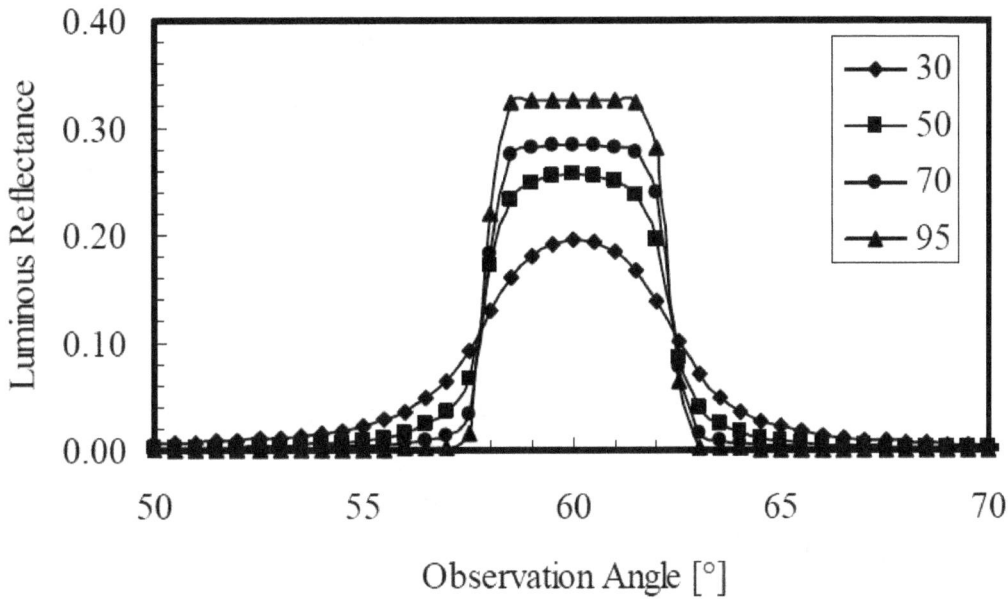

Figure 2

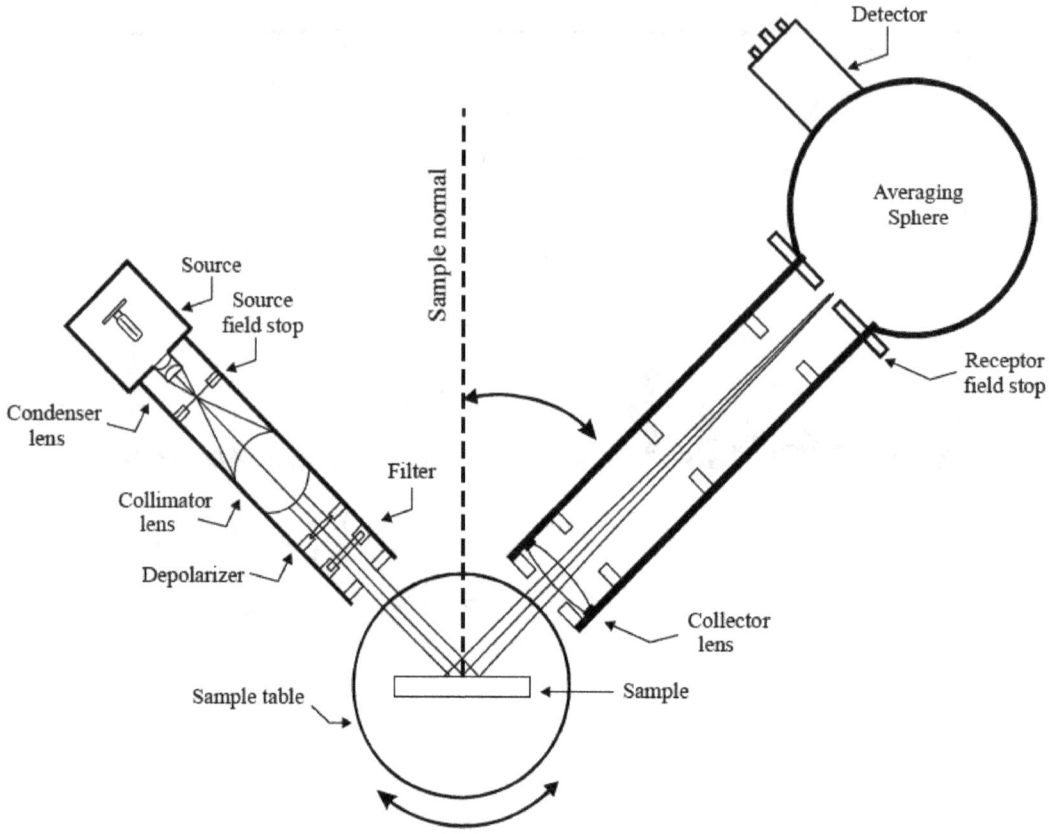

Figure 3

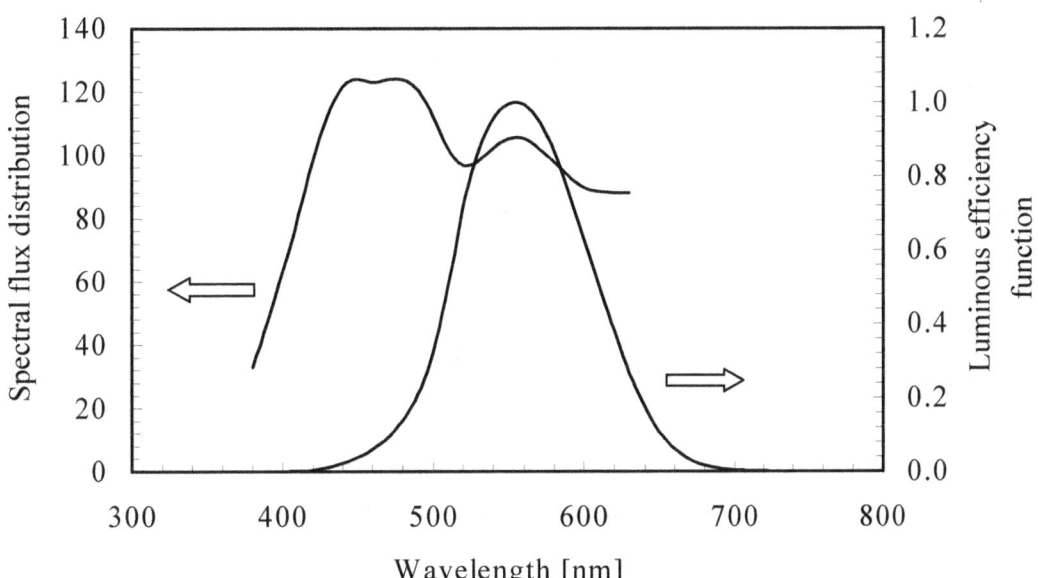

Figure 4

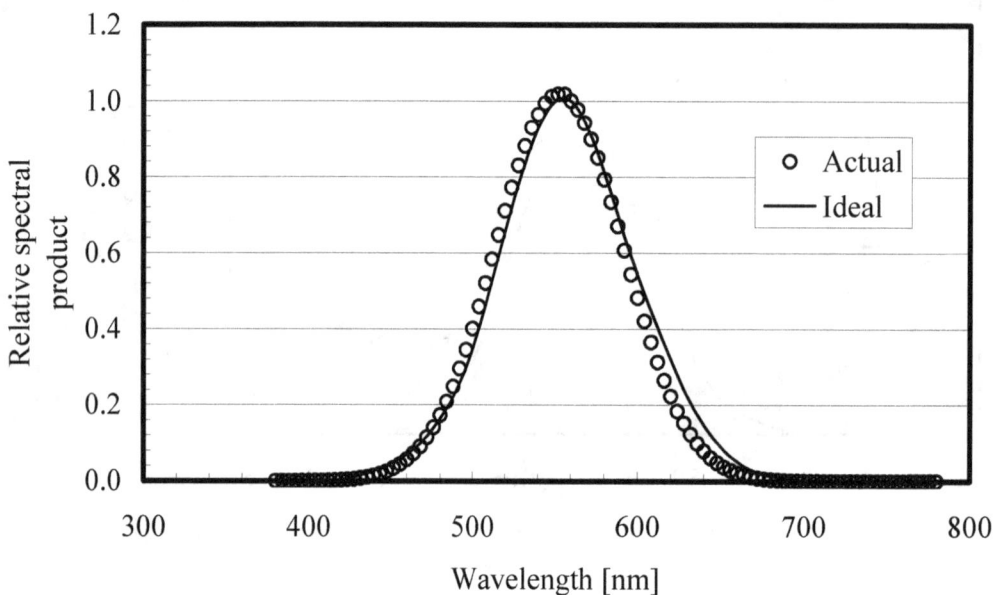

Figure 5

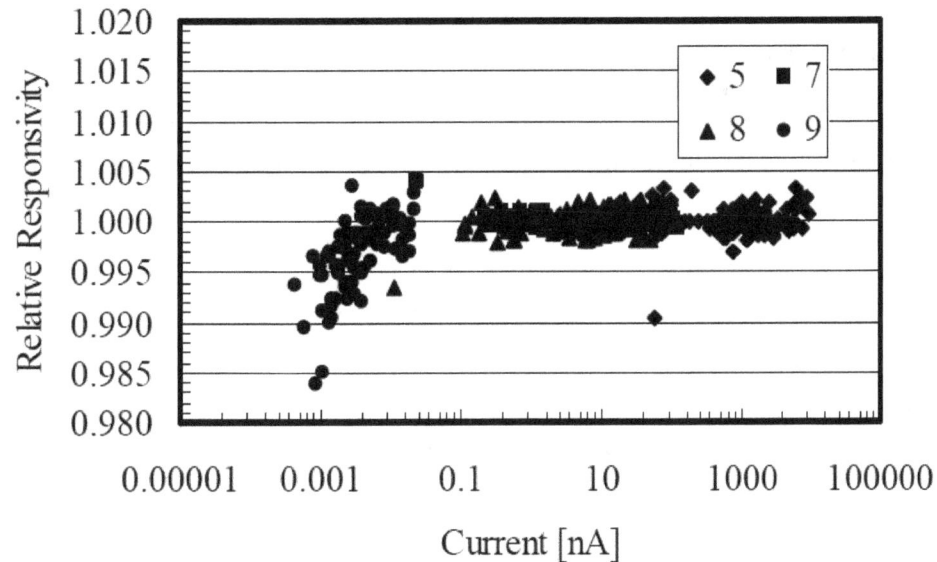

List of Tables

1. Standard geometries for specular gloss measurements and their applications.
2. Specular reflectance $\rho_0(\theta, \lambda_D)$ of the theoretical gloss standard for each incident angle of the standard geometries at wavelength $\lambda_D = 589.3$ nm.
3. ASTM specifications and the NIST measured values for source and receptor apertures, with tolerances and uncertainties, respectively.
4. Average gain ratio and relative standard deviation for successive amplifier gain ratios.
5. Correction factors for deviations from the standard recommendations of an unpolarized source, a perfectly collimated incident beam, nominal incident angle of 20°, 60°, and 85°, and spectral conditions.
6. a. Components of uncertainty that depend on the scattering properties of the materials and the resulting relative standard uncertainties. The values are based on a BaK50 primary gloss standard and a highly polished black glass test sample.
   b. Components of uncertainty that are independent of the scattering properties of the materials and the resulting relative standard uncertainties.
7. Relative expanded uncertainties of specular gloss measured by the NIST reference goniophotometer ($k = 2$) for the 20°, 60°, and 85° geometries.

Table 1

Specular Angle	Applications
20°	High gloss of plastic film, appliance and automotive finishes
30°	High gloss of image-reflecting surfaces
45°	Porcelain enamels and plastics
60°	All ranges of gloss for paint and plastics
75°	Coated waxes and paper
85°	Low gloss of flat matte paints and camouflage coatings

Table 2

Incident Angle	$\rho_0(\theta, \lambda_D)$
20°	0.049078
60°	0.100056
85°	0.619148

Table 3

Apertures	In plane (°)		Perpendicular to the plane (°)	
	ASTM Spec.	NIST Apertures	ASTM Spec.	NIST Apertures
Source	0.75 ± 0.25	0.75 ± 0.01	3.0(max.) ± 0.5	3.01 ± 0.01
20° receptor	1.8 ± 0.05	1.82 ± 0.01	3.6 ± 0.1	3.64 ± 0.02
60° receptor	4.4 ± 0.1	4.44 ± 0.02	11.7 ± 0.2	11.72 ± 0.01
85° receptor	4.0 ± 0.3	4.03 ± 0.01	6.0 ± 0.3	6.02 ± 0.01

Table 4

Gain ratio	Average value	Relative standard deviation [%]
G_6/G_5	9.999	0.002
G_7/G_6	9.998	0.002
G_8/G_7	9.997	0.002
G_9/G_8	10.017	0.003
G_{10}/G_9	9.952	0.003

Table 5

Deviation	Correction factors		
	20°	60°	85°
Source-polarizer, C_p	0.999890	1.000137	1.000222
Source-depolarizer, C_p	0.999999	1.000001	0.999995
Divergence, C_d	0.999999	0.999806	0.999999
Nominal angle, C_a	0.999996	0.999875	0.999940
Spectral Condition, C_s	1.000012	1.000006	0.999997

Table 6a

Component of Uncertainty	Effect	Type	Relative standard uncertainty [%]		
			20°	60°	85°
Source-polarizer	S	B	0.01	0.01	0.02
Source-depolarizer	S	B	0.0001	0.0001	0.0005
Source divergence	S	B	0.0001	0.02	0.0001
Nominal angle	S	B	0.0004	0.01	0.006
Spectral condition	S	B	0.001	0.0006	0.0003

Table 6b

Component of Uncertainty	Effect	Type	Relative standard uncertainty [%]		
			20°	60°	85°
Primary gloss standard	S	B	0.1	0.06	0.005
Source stability	R	A	0.01		
DVM	S	B	0.002		
Signal noise	R	A	0.01		
Photodiode linearity	S	A	0.09		
Amplifier gain ratio	S	A	0.002		
Repeatability	R	B	0.1		

Table 7

Geometry [°]	Relative expanded uncertainty [%]
20	0.34
60	0.30
85	0.27

Appendix C

New Primary Standard for Specular Gloss Measurements

María E. Nadal and E. Ambler Thompson
Optical Technology Division
National Institute of Standard and Technology
Gaithersburg, MD 20899

ABSTRACT

The measurement of specular gloss consists of comparing the luminous reflectance from a specimen to that from a calibrated gloss standard, under the same geometric conditions. The specifications for the gloss standard are discussed. A new primary gloss standard using BaK50 barium crown glass has been developed. Different calibration procedures are detailed, and the new standard is compared with other primary standards.

Keywords: specular gloss; gloss standard; glossmeter; goniophotometer.

I. Introduction

Specular gloss is the perception by an observer of the mirror-like appearance of a surface. This appearance cannot be measured; only the specific reflectance characteristics of the surface are measured. Gloss is a commercially important attribute of many materials such as paints, papers, plastics, and textiles, and is affected by the production, storage and use of these materials products. Several documentary standards describe the proper measurement conditions to determine specular gloss for specific surfaces. Particularly, the International Organization for Standards ISO 2813 [1] and the American Society for Testing Materials ASTM D523 [2] describe the measurement procedure that best correlates with visual ranking of specular gloss for nonmetallic samples. These standards specify the geometrical and spectral conditions of measurement. The incident beam, either collimated or converging, has a spectral flux distribution of Commission International de l'Eclairage (CIE) [3] Illuminant C and incident angles of 20°, 60°, or 85°. These angles are referred to as the standard geometries. The reflected beam is measured with a detector having the CIE luminous efficiency function V_λ. [3] Figure 1 shows a graphical representation of CIE illuminant C and the luminous efficiency function. The angular dimensions of the apertures for the source and detector are also specified for each incident angle.

For practical reasons, the specular gloss of a sample is measured relative to that of a primary standard, whose specular gloss is in turn determined relative to that of the theoretical standard [4]. This theoretical standard is specified to be a highly polished plane black glass with an index of refraction at the wavelength of the sodium D line, 589.3 nm, of $n_D = 1.567$. The specular reflectance of unpolarized light for this index of refraction is calculated using the Fresnel equations, and assigned a specular gloss value of 100 for each of the three standard geometries.

Since there is no black glass with exactly $n_D = 1.567$, national metrology institutes use primary standards to realize their scales of specular gloss. Two such primary standards have been in use, Carrara black glass [5] and quartz [6]. A new primary standard for specular gloss developed by the National Institute of Standards and Technology (NIST) using barium crown glass BaK50 [7] is reported here, which has several advantages over other primary standards. This new primary standard and the recently refurbished NIST reference goniophotometer [8] will provide accurate specular gloss measurements. The theory of specular gloss standards and measurements is presented first, followed by a discussion of other primary standard materials. The selection criteria for the new NIST primary standard are described, along with the index of refraction, specular gloss, and luminous reflectance properties of the standard. Finally, the new primary standard is compared to other standards.

II. Theory

Determining the specular gloss of a test sample involves measuring the luminous reflectance of the sample and a primary standard and determining the specular gloss value of the primary standard relative to that of the theoretical standard. The equations describing these measurements are presented in this section.

For unpolarized light, the specular reflectance ρ from the surface of a dielectric sample depends on the incident angle θ relative to the normal of the sample and the wavelength λ. The specular reflectance as a function of these variables is given by the Fresnel equations, namely

$$\rho(\theta, \lambda) = \frac{1}{2}\left[\frac{\sin^2(\theta-\theta')}{\sin^2(\theta+\theta')} + \frac{\tan^2(\theta-\theta')}{\tan^2(\theta+\theta')}\right], \qquad (1)$$

where θ' is the angle of refraction given by

$$\theta' = \arcsin(\sin(\theta)/n(\lambda)). \tag{2}$$

In general, the luminous flux Φ_v is given by

$$\Phi_v = K_m \cdot \int S(\lambda) \cdot V_\lambda(\lambda) \cdot d\lambda, \tag{3}$$

where $K_m = 683$ lm/W is the maximum spectral luminous efficacy for photopic vision, $S(\lambda)$ is the spectral flux distribution, and $V_\lambda(\lambda)$ is the spectral luminous efficiency function. The luminous flux incident on a sample Φ_i is given by Eq. (3), while the luminous flux reflected by a sample Φ_r is given by

$$\Phi_r = K_m \cdot \int S(\lambda) \cdot \rho(\theta,\lambda) \cdot V_\lambda(\lambda) \cdot d\lambda. \tag{4}$$

Therefore, the luminous reflectance $\rho_v(\theta)$ of a sample is given by the ratio of Eq. (4) to Eq. (3),

$$\rho_v(\theta) = \frac{\int S(\lambda) \cdot \rho(\theta,\lambda) \cdot V_\lambda(\lambda) \cdot d\lambda}{\int S(\lambda) \cdot V_\lambda(\lambda) \cdot d\lambda}. \tag{5}$$

The specular gloss of a sample under test, $G_t(\theta)$, is given by

$$G_t(\theta) = G_s(\theta) \cdot \frac{\rho_{v,t}(\theta)}{\rho_{v,s}(\theta)}, \tag{6}$$

where $G_s(\theta)$ is the specular gloss of the primary standard and $\rho_{v,t}(\theta)$ and $\rho_{v,s}(\theta)$ are the specular luminous reflectances of the test sample and primary standard, respectively. Furthermore, the specular gloss of the primary standard is given by

$$G_s(\theta) = G_0(\theta) \cdot \frac{\rho_s(\theta, \lambda_D)}{\rho_0(\theta, \lambda_D)}, \tag{7}$$

where $G_0(\theta)$ is the specular gloss of the theoretical standard and $\rho_s(\theta, \lambda_D)$ and $\rho_0(\theta, \lambda_D)$ are the specular reflectances of the primary and theoretical standards, respectively, at wavelength $\lambda_D = 589.3$ nm. Combining Eqs. (6) and (7), the specular gloss of the sample under test is given by

$$G_t(\theta) = G_0(\theta) \cdot \frac{\rho_{v,t}(\theta)}{\rho_{v,s}(\theta)} \cdot \frac{\rho_s(\theta, \lambda_D)}{\rho_0(\theta, \lambda_D)}. \tag{8}$$

From the documentary standards, the index of refraction of the theoretical standard at wavelength λ_D is $n(\lambda_D) = 1.567$, from which the specular reflectance $\rho_0(\theta, \lambda_D)$ is calculated using Eqs. (1) and (2). These specular reflectances are listed in Table 1 for the angles of incidence specified in the documentary standards. For each angle of incidence, the specular gloss of the theoretical standard is defined as $G_0(\theta) = 100$. Thus, for example, a specular reflectance $\rho_0(60°, \lambda_D) = 0.100056$ corresponds to a specular gloss value of 100. Similarly, the specular reflectance of the primary standard $\rho_s(\theta, \lambda_D)$ is determined from its $n(\lambda_D)$ using Eqs. (1) and (2). These values for the new primary standard are presented in Section IV. B.

At this point, three of the five quantities on the right-hand-side of Eq. (8) are known. The other two quantities are the luminous reflectances of the primary standard and test sample, which are generally not measured separately. Rather, the ratio of these reflectances is measured, given by the ratio of the luminous fluxes reflected from the test sample and primary standard, $\Phi_{r,t}$ and $\Phi_{r,s}$, respectively. Therefore,

$$\frac{\rho_{v,t}(\theta)}{\rho_{v,s}(\theta)} = \frac{\Phi_{r,t}(\theta)}{\Phi_{r,s}(\theta)} = \frac{\int S(\lambda) \cdot \rho_t(\theta,\lambda) \cdot V_\lambda(\lambda) \cdot d\lambda}{\int S(\lambda) \cdot \rho_s(\theta,\lambda) \cdot V_\lambda(\lambda) \cdot d\lambda}. \tag{9}$$

Using the definition for $G_0(\theta)$ and Eq. (9), Eq. (8) becomes

$$G_t(\theta) = 100 \cdot \frac{\Phi_{r,t}(\theta)}{\Phi_{r,s}(\theta)} \cdot \frac{\rho_s(\theta, \lambda_D)}{\rho_0(\theta, \lambda_D)} \, . \qquad (10)$$

This equation expresses the gloss value of the test sample as a function of the measured reflected luminous fluxes for the test sample and primary standard and the specular reflectances of the primary and theoretical standards.

From the discussion above, the specular gloss value of the primary standard is determined using monochromatic radiant flux, from Eqs. (1), (2), and (7), while the luminous flux reflected from the primary standard is used to calibrate the specular gloss of the test sample, from Eq. (10). This ambiguity produces differences in practical gloss measurements of up to 0.5 % since this recommendation ignores the dispersion characteristic of the material represented by changes in refractive index as a function of wavelength. It has been suggested in Ref [9] that this ambiguity be removed by defining the specular gloss of the primary standard from its luminous reflectance, given by Eq. (5). There are two methods for determining the luminous reflectance of the primary standard, one from measurements of the index of refraction over the wavelength range of visible light, the other from direct measurements of this reflectance. Results from both methods are presented below.

III. Other Primary Standards

The primary specular gloss standard previously used at NIST was a commercial Carrara black glass with $n_D = 1.527$ [5]. A recent evaluation of this type of glass and its constituents was performed, and no match to any commercially available optical glass could be determined. The National Research Council (NRC) of Canada investigated the stability of highly polished black glasses used as specular gloss standards [10]. They found that surface chemical contamination of these standards results in variations of 0.3 % to 0.5 % in the index of refraction over a time period of three to four years. These variations correspond to changes of about 2 % and 1 % in the specular gloss values for the 20° and 60° standard geometries, respectively. Repolishing the surface with cerium oxide recovered the original specular gloss values of these standards. They also investigated the uniformity of the index of refraction, which varied across the surface by approximately 0.5 %, and concluded that these variations were a result of inhomogeneities in the glass.

In general, there are a number of disadvantages to using black glass as a primary specular gloss standard, primarily inhomogeneity and aging of the glass, as well as easy damage to the front surface, requiring frequent repolishing and recalibration. The ISO 2813 suggests an alternative to black glass – a clear glass with roughened edges and back surface, with the back surface painted black to absorb any transmitted light. A difficulty with such a primary standard is finding a black paint that provides a good match to the index of refraction of the clear glass. Otherwise, some scattered light from the back surface will enter the detector and cause an error in the measured flux. An alternative to the black paint is to cut the clear glass at an angle so that back reflections are not incident on the detector, but this option is not feasible for small, portable instruments commonly used in industry. In 1980, NRC developed wedged quartz samples as their primary standards. Quartz was selected because of its high optical quality, stability, and durability. However, it does not have the same dispersion characteristics as the crown black glass standards used previously at NIST and commonly in industry.

IV. New Primary Standard
A. Description

The disadvantages of the previous primary standards for specular gloss described above motivated a search for a new primary standard. A good standard should be smooth, stable, cleanable, durable, uniform, and homogeneous with known optical properties. The selection criteria for the new primary standard were threefold. First, the standard would be a commercially available, high-purity optical glass with high chemical and mechanical durability. Second, the index of refraction n_D should be as close to 1.567 as possible to conform to the documentary standards. Third, the material should be homogenous and have dispersion characteristics similar to black glass.

An optical quality barium crown glass BaK50 was selected as the new NIST primary standard since it possesses high chemical and mechanical durability and n_D = 1.5677. In addition, this glass has a better homogeneity of the index of refraction than does black glass. The manufacturer's specifications are as follows: no bubbles, index of refraction homogeneity within 5 x 10^{-6} over an area of 70 mm^2. Three pieces of BaK50 with dimensions of 98 mm by 98 mm by 20 mm were purchased, and the NIST optical shop fabricated these pieces into samples with a 6° wedge at the back surface. This angle is sufficient to reflect the light incident on the back surface out of the field of view of the detector for all of the standard geometries, as shown in Figure 2. The front and back surfaces were polished to a roughness of 0.6 nm to 1.0 nm and the edges were roughened. A black felt material was placed at the back and edges of the samples to eliminate reflections from the surrounding black anodized holder.

B. Properties
1. Index of Refraction

According to the documentary standards, primary standards for specular gloss are calibrated from measurements of the index of refraction. The index of refraction of BaK50 was measured on a prism-shaped sample prepared from the same melt as the wedge samples. The minimum deviation technique [11] was used to determine the index of refraction at five different wavelengths with a standard uncertainty of 4 x 10^{-5}. The index of refraction as a function of wavelength is given in Table 2. The data was inter/extrapolated using a cubic spline to give the dispersion curve for wavelengths from 380 nm to 780 nm at 10 nm interval shown in Figure 3.

2. Specular Reflectance

Specular gloss values for the primary standard were determined from the index of refraction n_D, as specified in the documentary standards. From Table 2, n_D = 1.5677 at wavelength λ= 589.3 nm, and using Eq. (1) the specular reflectance for each standard geometry was calculated. These reflectances and their corresponding gloss values, from Eq. (7) and Table 1, are listed in Table 3.

3. Luminous Reflectance

The discussion at the end of Section II suggested that the specular gloss value of the primary standard be defined from its luminous reflectance, and noted that there are two methods for determining this reflectance. The first calculates the luminous reflectance from the index of refraction as a function of wavelength, while the second measures the luminous reflectance directly. The luminous reflectances and specular gloss values from these different methods are listed in Table 4, with the details of the calculations and measurements given below.

For the first method, the index of refraction as a function of wavelength, from Table 2 and Fig. 3, is used in Eqs. (1) and (2) to calculate $\rho_s(\theta, \lambda)$ for each of the standard geometries from 380 nm to 780 nm every 10 nm. The luminous reflectance $\rho_{v,s}(\theta)$ is calculated from Eq. (5) using $\rho_s(\theta, \lambda)$, CIE Illuminant C for $S(\lambda)$, and the CIE 1931 2° Observer for $V_\lambda(\lambda)$. Finally, the specular gloss value $G_s(\theta)$ for the primary standard is given by Eq. (7) but with $\rho_{v,s}(\theta)$ in place of $\rho_s(\theta, \lambda_D)$. The relative expanded uncertainty (k=2) for the luminous reflectance is <0.002 %.

The second method measures luminous reflectance by an absolute technique using the NIST reference goniophotometer [8]. For a fixed specular geometry, the following steps are followed for absolute luminous reflectance measurements. The sample is manually removed from the path of the incident beam, the detector is rotated into the beam path, and the net incident signal is measured. The sample is then placed into the beam path, the sample table and detector arm are rotated to the desired geometry, and the net reflected signal is measured. The net incident signal is measured again, and the two values are averaged to obtain the final incident signal. The luminous reflectance is calculated from the ratio of the net reflected signal to the net incident signal. From the measured luminous reflectance of the primary standard, the specular gloss value was calculated as described in the previous paragraph and these quantities are listed in Table 4. The listed values represent the average values for two NIST BaK50 samples. The third sample requires additional polishing. The 85° geometry was not measured due to limitations in the source aperture and the signal to noise level. The relative expanded uncertainty ($k=2$) for the luminous reflectance of BaK50 at the specular geometries of 20° and 60° is 0.22 %. The random effects, which include the source stability and detector noise, result in a relative expanded uncertainty ($k=2$) of 0.1 % for both geometries. The systematic effects, which include the DVM accuracy, amplifier gain, detector linearity, source polarization, angular scale and spectral product, result in a relative expanded uncertainty ($k=2$) of 0.2 % for both geometries.

Several important conclusions can be drawn from the results presented in Tables 3 and 4. The specular gloss value of this standard is nearly 100 for all three standard geometries, as shown in Table 3, a consequence of n_D of the new primary standard being close to the theoretical value of 1.567. Comparing Tables 3 and 4, the specular gloss values calculated from the luminous reflectance are greater than those calculated using only n_D for the 20° and 60° standard geometries, and are approximately equal for the 85° standard geometry. Finally, the luminous reflectances and corresponding specular gloss values obtained from each of the two methods, from Table 4, agree well with each other, the maximum difference in $G_s(\theta)$ being only 0.2. The uncertainty analysis shows that the calculated gloss value following the two different procedures agree within the uncertainties.

V. Comparisons of Specular Gloss Standards

The dispersion characteristics of BaK50 were compared to those of black glass and quartz gloss standards. Three different types of highly polished black glasses were investigated - the gloss standard previously used at NIST and two currently used in industry. Neither black glass or quartz are a good match for $n_D = 1.567$. The refractive indexes for the black glass samples were measured using an Abbe refractometer at three different wavelengths and fit with a cubic spline function to give the dispersion curves, from 380 nm to 780 nm at 10 nm intervals, shown in Fig. 4. The uncertainty of these measurements is 0.0005. The refractive indexes for quartz listed in Ref. [12] was inter/extrapolated using a cubic spline to give the dispersion curve for wavelengths from 380 nm to 780 nm at 10 nm interval shown in Figure 4.

The normalized dispersion curves of BaK50, black glass, and quartz standards are shown in Figure 5. The index of refraction is normalized at a wavelength of 560 nm and plotted as a function of wavelength. The average of the dispersion curves for the black glass samples is plotted. The relative dispersion characteristic of BaK50 closely resembles that of the black glass samples studied in this paper, while quartz has a different characteristic. The international and national standards define the specular gloss value for the theoretical and working standards based upon a single refractive index, n_D, but the instruments are specified for polychromatic radiation and broad spectral range detectors. This ambiguity leads to a situation where the gloss value of the sample under test depends on the dispersive characteristics of the secondary-working standard, such as black glass. This is particularly important for instruments whose spectral characteristics are in poor agreement with those specified by the documentary standards. Ideally, the calibration of a test sample should not be affected by the properties of the working standard.

VI. Conclusions

The new NIST primary standard for specular gloss has three important advantages over other materials used as primary standards. First, BaK50 is a commercially available high-purity glass with good homogeneity. Second, since n_D of BaK50 is close to the value of 1.567 specified for the theoretical standard, the specular gloss values for this glass differ from 100 by less than 0.2 for all three standard geometries. Third, the relative dispersion of BaK50 is similar to that of black glass, which is important for calibrating samples with instruments whose spectral characteristics are not in good agreement with those specified by the documentary standards. This new standard and the NIST reference goniophotometer provide a facility for specular gloss measurements of the best possible accuracy.

Acknowledgments

The authors wish to express special thanks to Edward Early for many useful discussions and the collaboration with the National Institute of Standards and Technology researchers who are working on the measurement science for optical reflectance and scattering project.

References

[1] International Standard ISO 2813, "Paint and Varnishes-Measurements of Specular Gloss of Nonmetallic Paint Films at 20°, 60°, and 85°" (International Organization for Standardization 1978).

[2] "Standard Test Method for Specular Gloss, ASTM D532," American Society for Testing and Materials, West Conshohocken, PA (1995).

[3] Commission International de l'Eclairage: International Lighting Vocabulary, CIE Publ. No. 15.2 (1987).

[4] R. S. Hunter and R. W. Harold, The Measurement of Appearance, 2nd Ed., John Wiley & Sons, New York, N. Y. (1987).

[5] J. J. Hsia, "The NBS 20-, 60-, and 85- Degree Specular Gloss Scales," NIST Tech. Note 594-10 (1975).

[6] J. Zwinkels and M. Nöel, "Specular Gloss Measurement Services at the National Research Council of Canada," Surface Coatings International **12**, 512 (1995).

[7] Certain commercial equipment, instruments or materials are identified in this paper to specify adequately the experimental procedure. Such identification does not imply recommendation or endorsement by the National Institute of Standards and Technology, nor does it imply that the materials or equipment identified are necessarily the best available for the purpose.

[8] M. Nadal and A. Thompson, "NIST Reference Goniophotometer for Geometrical Appearance Measurements," submitted to the Journal of Coatings Technology.

[9] B. Wolfang and C.X. Dodd, "Stability Problems in Gloss Measurements," Journal of Coating Tech. **552**, 44-48, (1980).

[10] W. Budde, "The Calibration of Gloss Reference Standards," Metrologia **16**, 89–93 (1980).

[11] G. E. Fishter, "Refractometry," in the Applied Optics and Optical Engineering, Vol. IV, edited by R. Kingslake (Academic Press, New York, 1967), p. 363-382.

[12] I. H. Malitson, "Interspecimen Comparison of the Refractive Index of Fused Silica," Journal of the Optical Society of America **55**, 1205-1209 (1965).

List of Tables

Table 1   Specular reflectance $\rho_0(\theta, \lambda_D)$ of the theoretical gloss standard for each standard geometry

Table 2   Index of refraction n of BaK50 glass as a function of wavelength

Table 3   Calculated specular reflectance $\rho_s(\theta, \lambda_D)$ and specular gloss value $G_s(\theta)$ of the new primary standard at $\lambda_D = 589.3$ nm for each standard geometry

Table 4   Average luminous reflectance $\rho_{v,s}(\theta)$ and specular gloss value $G_s(\theta)$ for the New Primary Standard for each standard geometry and method of calculation or measurement.

Table 1

Standard Geometry	$\rho_0(\theta, \lambda_D)$
20°	0.049078
60°	0.100056
85°	0.619148

Table 2

Wavelength [nm]	n
435.8	1.5800
480.0	1.5753
546.1	1.5702
589.3	1.5677
643.9	1.5654

Table 3

Standard Geometry	$\rho_s(\theta, \lambda_D)$	$G_s(\theta)$
20°	0.049172	100.19
60°	0.100167	100.11
85°	0.619204	100.01

Table 4

| Method | Standard Geometry | | | | | |
| | 20° | | 60° | | 85° | |
	$\rho_{v,s}(\theta)$	$G_s(\theta)$	$\rho_{v,s}(\theta)$	$G_s(\theta)$	$\rho_{v,s}(\theta)$	$G_s(\theta)$
Index of refraction	0.049464	100.8	0.100510	100.5	0.619375	100.0
Luminous reflectance	0.049587	101.0	0.100743	100.7		

List of Figures

Figure 1	Graphical representation of the spectral power distribution of the CIE Illuminant C and the luminous efficiency function.
Figure 4.1	Schematic diagram of the new NIST primary standard showing the incoming and reflected beam at 60° specular geometry. The wedge deflects the reflection from the back surface away from the detector.
Figure 3	Fitted refractive index, $n(\lambda)$, for new primary standard as a function of wavelength
Figure 4	Fitted refractive index, $n(\lambda)$, for Carrara black glass, quartz, and black glass working standards as a function of wavelength.
Figure 5	Normalized index of refraction as a function of wavelength for the samples indicated in the legend.

Figure 1

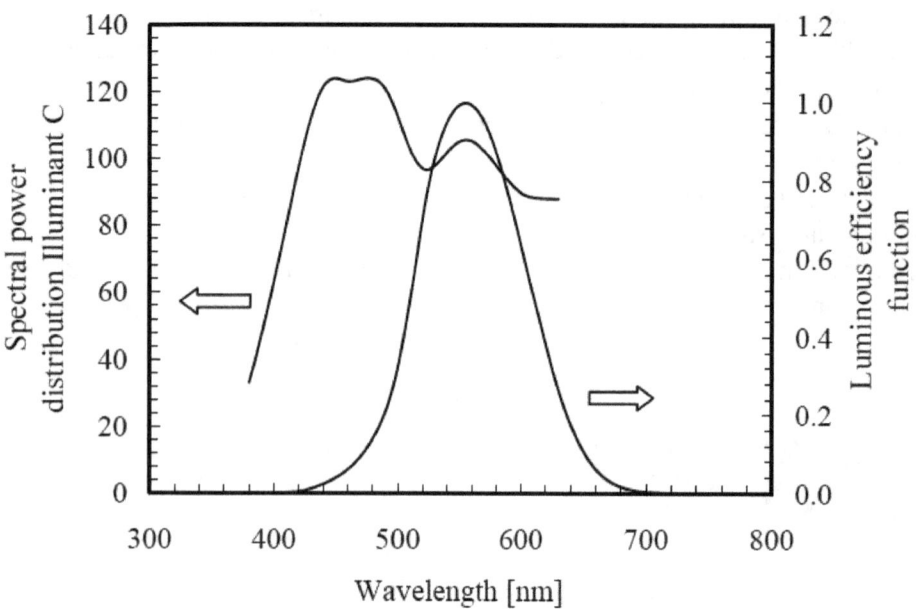

Figure 2

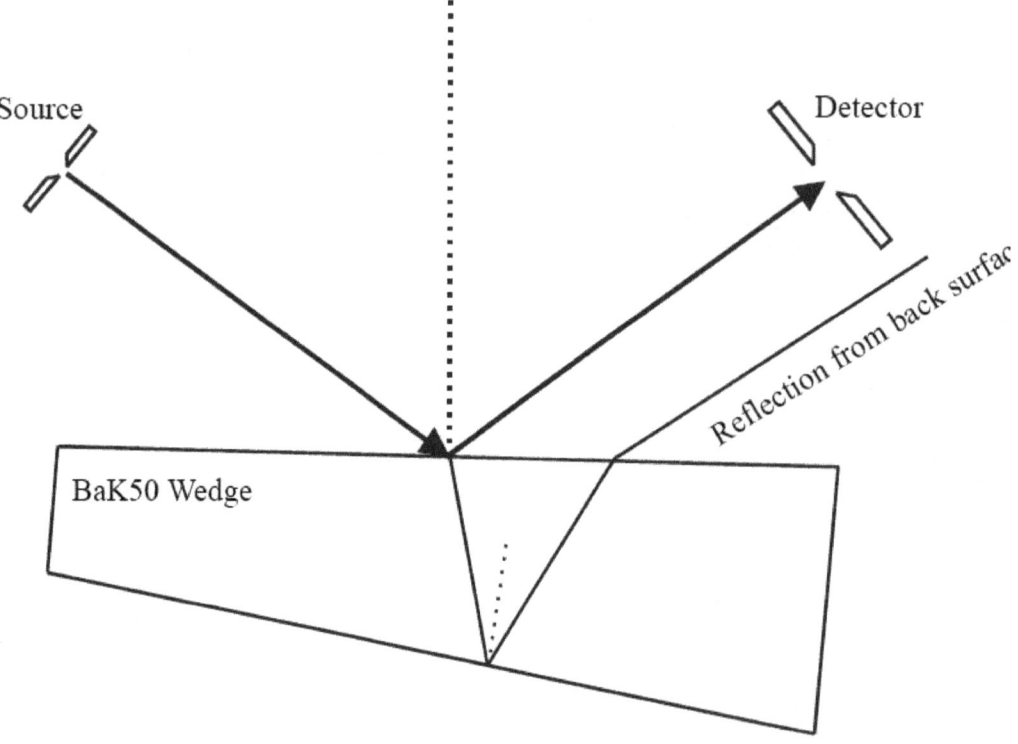

Figure 3

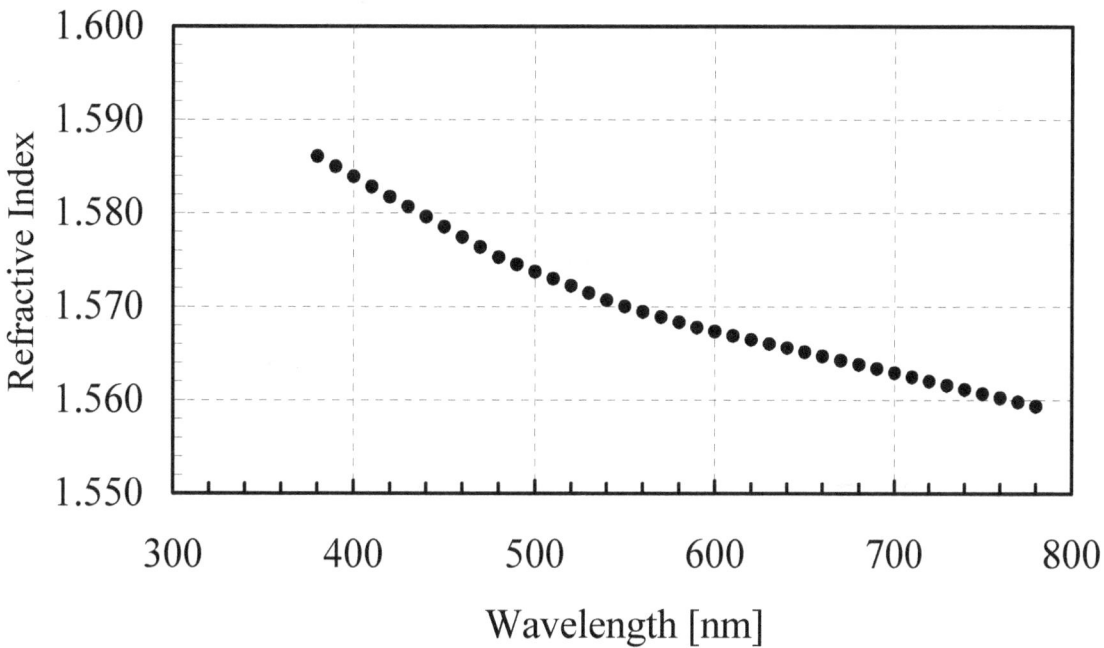

Figure 4

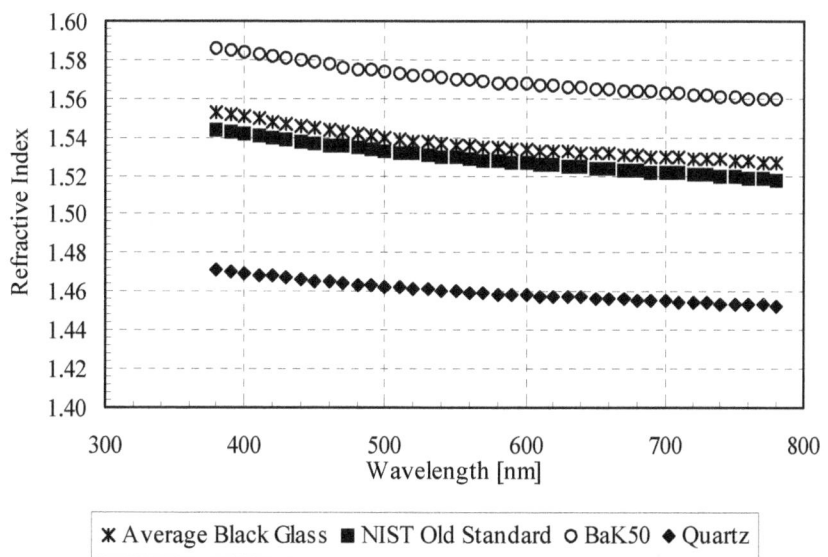

Figure 5

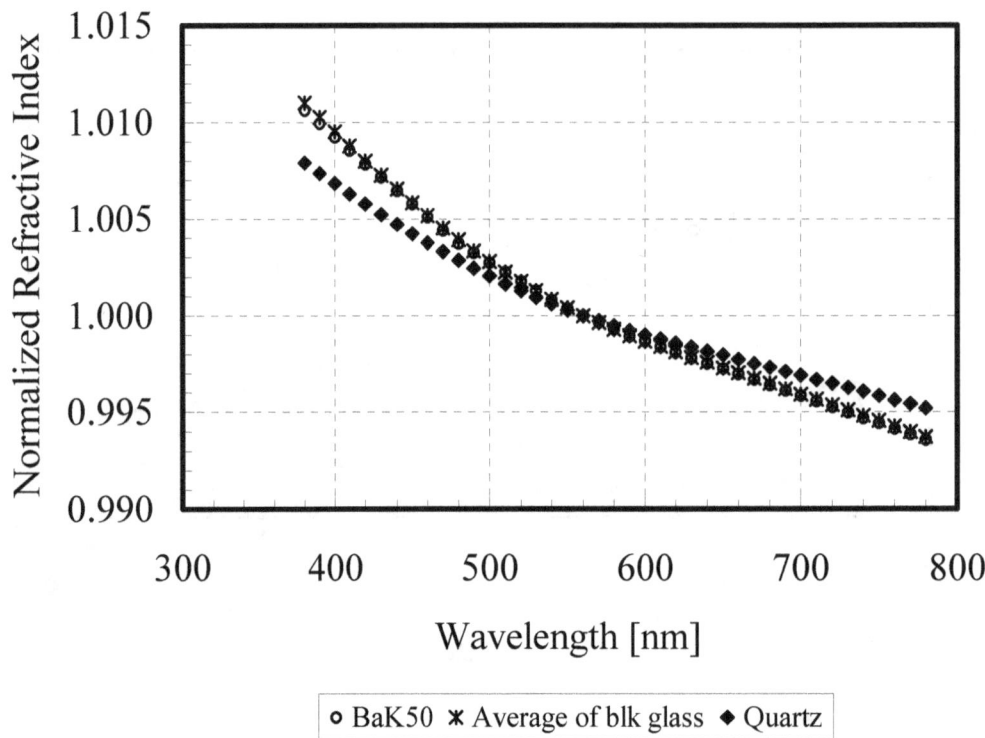

www.ingramcontent.com/pod-product-compliance
Lightning Source LLC
Chambersburg PA
CBHW081848170526
45167CB00007B/2926